Victor Vaughan

ALSO BY RICHARD ADLER

Cholera in Detroit: A History (McFarland, 2013)

Mack, McGraw and the 1913 Baseball Season (McFarland, 2008)

Victor Vaughan

*A Biography of the
Pioneering Bacteriologist,
1851–1929*

RICHARD ADLER

McFarland & Company, Inc., Publishers
Jefferson, North Carolina

LIBRARY OF CONGRESS CATALOGUING-IN-PUBLICATION DATA

Adler, Rich, author.
　Victor Vaughan : a biography of the pioneering bacteriologist, 1851–1929 / Richard Adler.
　　p.　　cm.
　Includes bibliographical references and index.

　　ISBN 978-0-7864-9599-3 (softcover : acid free paper) ♾
　　ISBN 978-1-4766-1784-8 (ebook)

　I. Title.
　[DNLM: 1. Vaughan, Victor C. (Victor Clarence), 1851–1929. 2. Physicians—Michigan—Biography.　3. Bacteriology—history—Michigan.　WZ 100]
　　QR31.A3　　　579.3092—dc23　　　[B]　　　2014043163

BRITISH LIBRARY CATALOGUING DATA ARE AVAILABLE

© 2015 Richard Adler. All rights reserved

No part of this book may be reproduced or transmitted in any form or by any means, electronic or mechanical, including photocopying or recording, or by any information storage and retrieval system, without permission in writing from the publisher.

On the cover: Dr. Victor Vaughan in his laboratory (Bentley Historical Library, University of Michigan); (inset) bacteria culture (Iceclanl, Wikimedia Commons)

Printed in the United States of America

McFarland & Company, Inc., Publishers
　Box 611, Jefferson, North Carolina 28640
　　www.mcfarlandpub.com

Table of Contents

Acknowledgments vi
Preface 1
1. Origin and Evolution of the Medical School at the University 7
2. Victor Clarence Vaughan: The Early Years 23
3. Member of the Faculty 29
4. Dean of Medicine 43
5. Ptomaines and Leucomaines 56
6. War and Disease 66
7. The Interim, 1898–1916 89
8. The 1910s: Dean and National Service Again 125
9. Influenza and the Great War 134
10. Retirement 179

Chapter Notes 191
Bibliography 203
Index 207

Acknowledgments

I would first like to express my appreciation to the staff of the Bentley Historical Library in Ann Arbor for their patience in helping find the location of images and files relevant to the story. The Dearborn campus Office of Research and Sponsored Programs and its director, Dr. Drew Buchanan, were generous in providing funds for purchase of these and other images and related material. I would also like to acknowledge the work of my student, Elise Mara, who possesses an ability, almost unique among undergraduate students, to locate obscure reference material and the patience to do so.

Preface

Though born and raised in a small town in Missouri, Victor Vaughan spent nearly all of his professional life in the college town of Ann Arbor, Michigan. It was here that he earned his three advanced academic degrees: a master of science (1875), doctorate (1876) and a medical degree (1878), as well as an honorary LL.D. (1900), one of four honorary doctor of laws degrees he would receive; he would have additional honorary degrees conferred by numerous other institutions. On the state level, Vaughan was a member of the Michigan State Board of Health for nearly three decades and was elected president during many of those years. On a national level, Vaughan was elected president by both the Association of American Physicians (1908) and by the American Medical Association (AMA) (1914); he served as a member of the House of Delegates, the policy making body for the AMA, from 1902 to 1904 and again in 1906. These were only a sampling of the state and national organizations with which Vaughan was associated during his long career.

But it was during his long tenure at the University of Michigan that Vaughan had his greatest impact. After earning an undergraduate degree at Mount Pleasant College in Missouri in 1872, and teaching Latin and chemistry there for another two years, Vaughan entered the University of Michigan. His decision to enroll there was fortuitous. Residents of a border state during the Civil War, Missourians were split in their support for either the Union or the Confederacy; Vaughan's family had been Confederate sympathizers, which meant Vaughan would not be allowed to teach in Missouri's public schools. That, and his strong interest in chemistry, led to his applying to Michigan.

Vaughan's first job with the university was that of a teaching assistant in chemistry. After earning a master's degree, Vaughan continued with his graduate work in the doctoral program, producing two theses and receiving his Ph.D. in 1876. He continued at Michigan in the medical school, earning his

M.D. in 1878. During these years he quickly rose in the teaching ranks: assistant in chemistry (1875) to instructor in medicinal chemistry and lecturer in physiology (1877), lecturer in medicinal chemistry (1879), assistant professor in medicinal chemistry (1880) and full professor of physiological and pathological chemistry as well as associate professor of therapeutics and *materia medica* in 1883.

As a full member of the faculty in the 1880s, Vaughan established the first hygienic laboratory. This served as the state health laboratory well into the following century, carrying out much of the diagnostic work for the state during the next two decades.

Vaughan maintained a medical practice for many years, but it was in the fields of research and teaching, particularly that of physiological chemistry and its application in the new science of bacteriology, that he had his greatest success. Bacteriology had its beginnings at Michigan during the early 1880s with a course known as Sanitary Science. Vaughan, still working as an assistant professor, was the instructor. Topics included disease germs, physiological ferments and vaccinations. The contents of these courses were new in those early years of bacteriology. The "germ theory of disease"—the concept that disease was not the result of emanations from putrefying soil (miasmas), but was the result of an infectious agent, a germ—was just beginning to develop.

The emphasis on the role of bacteria—germs—and disease was quickly applied by Vaughan in the growing field of public health. As it became increasingly apparent that disease often resulted from bacterial contamination of food or water supplies, Vaughan began investigations into outbreaks of poisoning among consumers of milk and cheese products in the state. Vaughan isolated and identified a poison he called tyrotoxicon, determining that its origin was the result of bacterial putrefaction of stored milk.

The lack of an efficient hygienic laboratory increased the difficulty in solving problems which affected public health, and in 1884 Vaughan began to lobby the state board of health to finance the building of such a laboratory. After several years Vaughan's arguments proved successful, and the Hygienic Laboratory was built and functional before the end of the decade; Vaughan became its director.

The field of bacteriology being new during these years, few laboratories existed in the United States where training in bacteriological techniques could be carried out. Taking advantage of the scientific expertise which existed on the European continent, during the summer of 1888 Vaughan traveled to Germany where he had the opportunity to learn techniques firsthand in the laboratory of Robert Koch. Koch's associate Carl Fraenkel provided much of the instruction for Vaughan and his associate from Michigan who was traveling with him, Frederick Novy. The collaboration between Vaughan and Novy would continue for decades even after Novy himself became a prominent member of the scientific community.

In 1891 Vaughan became dean of the University of Michigan Medical School, a position he would hold for three decades. His contributions, both administrative and scientific, would continue to have an impact on the university long after his retirement. While the medical school had grown significantly since its beginnings in 1850 and its standards were among the highest in the nation—arguably only those at Johns Hopkins were stronger—Vaughan quickly implemented the changes which had evolved from his own interest in laboratory work and research. Older faculty, some set in their ways, were eased out over time and replaced by those with strong interests in both teaching and research. Perhaps this emphasis on clinical and laboratory research was among the most significant areas of modernization instituted by Vaughan during this period.

Vaughan continued his own work, applying his interests in physiological chemistry to the evolving field of medical bacteriology. Beginning with his collaboration with Novy in the 1880s, he developed his own theories as to the role of intoxication as an etiological basis of disease; his first love, chemistry, was always the underlying basis for this work. His experiments on the formations of toxic chemicals in cells, including those resulting from the actions of pathogenic bacteria, led to his theories on the role of ptomaines and leucomaines as metabolic by-products as well as the formation of chemical poisons due to the splitting of proteids (proteins). Some of his ideas of course became dated as more scientific knowledge became available; for example his arguments concerning anaphylaxis and the basis for adaptive immunity sometimes differed significantly from those coming out of the German laboratories. But his theories were always directly based upon the data generated from his scientific experiments; few contemporaries found fault with his arguments. Vaughan's résumé ultimately contained over two hundred publications, scientific papers as well as books. His legacy also included five sons, some of whom followed him in research endeavors. One of his sons, Henry Frieze Vaughan, was the recipient of the first doctor of public health awarded by the university. After serving in World War I as a captain in the Sanitary Corps, following in the footsteps of his father, Henry Vaughan was appointed commissioner of the Department of Public Health in Detroit. He held this position until 1941 when he became dean of the School of Public Health at the university.

Twice Victor Vaughan was called upon to serve his country in time of war. During the Spanish-American War (1898) he volunteered for service and was posted in Cuba. While there he briefly came under fire, and while tending the injured in his role as a physician he was exposed to yellow fever, a mosquito-borne viral disease; he became seriously ill and barely survived. In response to the high proportion of water-borne illnesses among the troops, principally typhoid fever, Vaughan was appointed to a commission which included Drs. Walter Reed and Edward Shakespeare. The commission spent weeks touring military camps in the southeastern United States, observing the deplorable sanitary conditions which were the direct source of the typhoid outbreaks among the troops. Both Reed and Shakespeare died before the report from the commission was completed, and it was left to Vaughan as the surviving member to finish the work. The report, published in 1904, described in detail the breakdown in sewage disposal in the camps, a combination of both incompetence on behalf of the medical service and the inability of administrative services to deal with the large influx of men in a short period of time. Vaughan's recommendations resulted in the emphasis on proper sanitation and sewage disposal the military would implement in the future.

During World War I, or the Great War as it was called by its contemporaries, Vaughan was appointed to the National Research Council. Vaughan's duties included overseeing sanitation and hygiene in the army camps, emphasizing the importance of dealing with infectious disease. A medical disaster beyond Vaughan's control appeared during the last year of the war: influenza. The virulence of the disease surprised even Vaughan, and beyond attempts at quarantine there was little anyone could do to limit the spread.

Vaughan retired from his duties as an administrator at Michigan in 1921, but his professional work continued nearly until his death in 1929. He continued to publish even as late as 1927, producing a report for the American Chemical Society that year in which he included a summary of decades of work with bacteria; ill health prevented Vaughan from attending and the report was read by his son.[1] That previous fall of 1926 he had traveled to China, Japan and the Philippines on behalf of the Medical Congress, a trip which may have contributed to the failing health in a man well into his seventies by then. Vaughan's legacy continues at the university more than eight decades after his death. More important than the eponymous structures which do or did exist at the university is the Victor Vaughan Student Society. Established in 1929, it continues today as a mechanism for students to develop or express their interest in medicine through papers and presentations on that subject.

But while this book is the story of a man and his contributions to science in general, and bacteriology more specifically, it is meant to be more than just that. Victor Vaughan himself completed an autobiography during his retirement years, *A Doctor's Memories* (1926), with which anyone interested in his personal life and career may find enjoyment; the autobiography was immensely helpful as a guide in completing this story. An overview of the history of the Michigan Medical School can be found in a number of sources; I recommend the late Horace Davenport's *Not Just Any Medical School* (1999), one of many sources I found particularly helpful, for anyone interested in the story of Michigan medicine from its beginnings until the 1980s. While I am far too young to have known the principals from the early years, I did have the opportunity to know some of the more recent individuals named by Davenport during and after my two-year employment as a research associate in human genetics during the tenure of Dr. James Neel. I can say firsthand that the strong tradition of a quality research environment in the medical school continues.

My intended audience is not only those interested specifically in Vaughan or the medical school per se. I have attempted to draw upon my interest in the history of bacteriology, microbiology as the field is now known, to provide contemporary accounts of the thinking and research carried out during the years of the story. The terminology and concepts learned often by rote among today's students are described directly in the words written by those who first observed or studied those concepts. I have intentionally quoted heavily from those accounts in order that today's readers can follow that original thinking (and even why terms were coined).

The story discussing the impact of Vaughan's career can be divided among several categories and topics. First is the story of the medical school itself. Let me acknowledge the importance of Davenport's work in this.[2] Davenport provided a fascinating overview of the early years of the school. His sources were helpful in tracking firsthand accounts. Using Davenport as a guide, this book attempts to cover in significantly more depth the story of the school, and emphasize the careers and lives of those who preceded Vaughan as dean. As was done throughout this book, I have utilized extensively quotes from original sources in depicting the story.

One cannot describe Vaughan's medical career without looking at the man himself. The second chapter describes his early life and education, explaining why he chose Michigan to continue the career path in chemistry he began as an undergraduate at Mount Pleasant College. His autobiography, *A Doctor's Memories,* was particularly important in this as well as in understanding later portions of his professional career.[3] The story continues with his appointment to the medical school faculty, his advancement in that capacity and ultimately to his appointment as dean of the University of Michigan Medical School. During these years he had the opportunity to carry out significant research in the study of pathogenic or toxigenic mechanisms, some of which is outlined above, and most of which is discussed in detail within the book. He also had the opportunity to collaborate or mentor numerous young scientists who ultimately became prominent in their own respective areas; Frederick Novy was arguably one of the most gifted and important of these individuals. Descriptions of some of these individuals, their backgrounds and works, are included.

Vaughan had the opportunity to investigate two significant epidemics during his years as dean: the typhoid epidemic among army troops during the Spanish-American War, and the epidemic-pandemic of influenza in 1918. I believe it helpful to study the history of each as they developed during the course of human civilization; Vaughan himself often described the importance of knowing the history of a disease when trying to understand its impact. Typhoid and influenza can likely be tracked well over two millennia, though the contemporary descriptions might apply to other illnesses as well. As examples, the Plague of Athens vividly described by the Greek historian Thucydides in his history of the Peloponnesian War was likely typhoid; the "fever" which killed Alexander the Great was probably not. I have included these histories, not only to observe the role played by Vaughan during his investigations of these outbreaks, but to allow the student of medical history to understand their impact on earlier populations. Extensive descriptions in the form of quotes from earlier observers are provided in this regard. I have also quoted extensively from Vaughan's contemporaries, providing context for his work. Descriptions of Vaughan's life and career may also be found in the numerous obituaries which rightly praised the man.[4]

Chapter 1

Origin and Evolution of the Medical School at the University

The medical school at the University of Michigan prior to the appointment of Victor Vaughan in the 1890s bore little relation to the school as it evolved after 1891. While it could still be included in the upper tier of such schools during those early years, the education of future physicians still consisted largely of lectures in a relatively few areas, with limited clinical or hands-on training for the students. Certainly there was almost no medical or bacteriological research being carried out; few opportunities for such research existed anywhere during this period in the United States. The duties of faculty consisted almost entirely of teaching or administrative activities. If a member of the faculty wished to carry out research, and few did, they did so in their spare time. But this did not mean that the University of Michigan could not be considered progressive in other areas of the medical field when compared with its contemporaries. Women had already been admitted and graduated from the medical program by the time Vaughan was appointed as dean. The first African American graduated in 1872. Before any discussion of Vaughan's tenure as dean of the medical school, and his subsequent role in medical history, it is helpful to briefly describe his predecessors in that role, and the origin and development of medical sciences at the university as well as the medical school itself, pre–1890.

The first official board of regents for the university was appointed by Michigan Governor Stevens Mason in 1837; among those appointees was at least one prominent Detroit physician, Dr. Zina Pitcher. Among the first acts of the board was the appointment of a scientist, Dr. Asa Gray, as professor of botany and zoology. Although Gray never carried out any form of instruction at the university, he was requested by the board to travel to Europe to buy books (and a herbarium) for the university; these purchases would later evolve into the first library at the university. Gray resigned from the university in 1842, accepting a position at Harvard University, where he would eventually become the most important botanist during the nineteenth century. Gray was succeeded at Michigan by Dr. Abram Sager (1810–1877), who had been serving as head of the botanical and zoological department for the Michigan State Geological Survey since 1837. Sager's background mirrored that of many of his contemporaries at the university. He had been educated at the Rensselaer Polytechnic Institute in New York State, where he had the opportunity to study with several of the most important botanists and zoologists of the time. After graduating in 1831 he continued with medical studies at Albany Medical School in New York and in Castleton, Vermont, from which he received his medical degree in 1835. He arrived in Detroit that year, establishing a medical practice which continued until his appointment to the state geological survey.

The second scientific appointment at the university was that of Dr. Douglass Houghton in 1839, named professor of chemistry, geology and mineralogy. Houghton was head of the Michigan State Geological Survey immediately prior to his appointment. As was the case with Sager, Houghton received his initial education at the Rensselaer Polytechnic Institute, graduating in 1831; shortly afterwards he obtained his license to practice medicine. As had been the situation with Asa Gray, Houghton never carried out any instruction but instead served in developing collections for the university. Houghton tragically drowned in 1845 while carrying out a geological survey in northern Michigan. Dr. Silas Douglas, appointed to be Houghton's assistant in 1844, succeeded Houghton.

The arguments in favor of the inception of the medical school at the university have much in common with the modern concern of "brain drain": the fear that trained professionals were leaving the state to study and seek jobs elsewhere. During the meeting of the regents in January 1847, Drs. Silas Douglas and Abram Sager, as well as other local physicians, "presented a 'memorial' or 'communication' to the board asking that a medical department be established at the university.... An informal survey had suggested that at least seventy Michigan residents had been forced out of state to get a medical education. Some of them were studying at schools that offered very poor training. Other state residents had not bothered to get any training at all and simply started to practice medicine."[1] The clear implication of the request was that students interested in medicine would enroll somewhere other than in Michigan, and once they were licensed would remain there. Even worse, some medical "practitioners" would be so poorly trained that they would represent a threat to the health of their patients.

Even granted that the concerns were legitimate, the nascent university was in no financial position to simply establish a new medical program in Michigan. The question had already been brought up earlier. In 1838 the regents had agreed that a college of surgeons and physicians should be established, at the time to be located in Detroit, but nothing came of that plan.[2] This time, however, the petitioners provided a possible solution to the financial difficulty. The program would consist of only four teaching faculty, covering the areas of anatomy, surgery, *materia medica* (therapeutic properties of medicine) and the practice of medicine. Fees collected from the students would be used to pay faculty salaries.[3] Space was also a problem and would need to be addressed. In 1847 the university was housed largely in a single building which contained not only the classrooms and teaching facilities, but also housing for the students.

After briefly assigning the question to a committee, the regents attempted to produce a compromise. They approved the request for an additional university building, but turned down the petition to create an actual medical department within the university. In response to the defeat, Pitcher moved that "the board deeming it expedient to proceed at an early day to the organization of the medical department of the university will at the next annual meeting [August 1847] appoint the requisite professors."[4] Pitcher's motion was defeated. Rather than simply accepting defeat, Pitcher brought the proposal again before the regents at their next meeting that August. This time the board agreed to establish another committee, headed by Pitcher and including board members Elon Farnsworth and Edward Mundy, to again determine the expediency of a new medical program and provide a report by the January 1848 meeting.[5]

The importance of such a program became clear in the report provided by Pitcher and the members of his committee. Pitcher by the time of the 1848 regents' meeting was a member of the recently established American Medical Association (AMA), founded in May 1847 in Philadelphia, Pennsylvania. Dr. Nathaniel Chapman had been elected as the first president of the AMA at the 1847 meeting and worked closely with Pitcher in providing the background for the latter's arguments.

Pitcher's membership with the AMA included participation in the association's education committee. The association's 300 members had met in Philadelphia that May. Among the topics discussed by the education committee was that of the poor state of medical education in the country. Consequently Pitcher was well informed about the status of medical schools in the nation. Many of the schools which did exist were merely "diploma mills," where for a fee, a student, including those from Michigan, could become licensed to practice. In support of his earlier argument dealing with establishment of a medical school at the university, Pitcher again placed in front of the Regents the situation elsewhere:

> It was stated [January 1847 meeting of the regents] that there were at that time seventy young men absent from the state in attendance upon the lectures of the professors at the several Medical Colleges south and west of us. The catalogues of these schools have not reached us for the present session of 1847–8, but from private letters we are informed that there are thirty Michigan students at Cleveland and ten at Chicago. The writer of the letter from Chicago remarks: "There are more students from Michigan at La Porte than at this place." Since the above was written a friend has placed in the hands of the committee the catalogue of the Western Reserve College, Ohio [now Case Western], from which it appears there are thirty-three students from this state at Hudson, making the ascertained number eighty or eighty-five. There are undoubtedly enough in attendance at Buffalo, Geneva, New York City, and Pennsylvania to make that whole number at least one hundred. Probably half that number are pursuing their studies in the offices of private physicians.
>
> From this state of facts your committee are of the opinion that in two years from the establishment of a medical department of the university there would at least fifty students matriculate annually, the number of course constantly increasing. Their initiation fees would furnish a fund sufficient for the gradual increase of the medical library.... Your committee on taking a general view of the profession of medicine in this state with the present prospect of an influx of half educated men from the schools about us, of the state of that profession at large in connection with the measures now adopted for its improvement, and the consequent amelioration of the physical condition of those who resort to such feeble instrumentalities for relief from corporal suffering, are led to the conclusion that it is our duty at once to proceed to carry into effect the objects embraced in the resolution submitted to their consideration.[6]
>
> The regents had a duty to the citizens of Michigan to change this situation. They had a duty not only to establish a medical department, as had been provided in the University Act of 1837, but "to establish a medical department of the university ... worthy of imitation."[7]

Specifically, Pitcher argued, "If this board, now that it has the power, will conform its actions to the views of the national medical convention [American Medical Association] by requiring a better preparation and a more extended range of professional study from those who graduate at the university ... your committee can anticipate the day when Michigan will be distinguished by the genius and skill of her sons in each of the learned professions."[8] The board of regents was convinced and voted unanimously to approve the establishment of a medical program.

Since the facilities were inadequate, the regents also approved an appropriation of $3,000 for construction of a building to house the medical program. Five members of the faculty were approved and appointed: "Douglass (pharmacy and medical jurisprudence), Sager (physic or medicine and obstetrics and the diseases of women and children), [Dr.] Moses Gunn (anatomy and surgery), [Dr.] Jonathan Adams Allen, Jr. (pathology and physiology), and [Dr.] Samuel Denton (physic)."[9] The new medical school was unique among those elsewhere around the country for several reasons. The salaries of the faculty were paid by the university, the first such example in the nation, and students were admitted free to the program. Unlike the situation with most other contemporary medical schools, students did not have to purchase admission tickets from the faculty in order to attend lectures or demonstrations.

The official founding date for the University of Michigan Medical School was therefore 1848; at least that was the year the regents commenced appointing the faculty. It would be two years before the appointed faculty began the actual instruction of medical students.[10] The medical facility was a three-story building then situated on what was the edge of campus; the building also containing the offices of the five members of the medical faculty. Professor Abram Sager, the replacement for Asa Gray, was chosen as the first dean. His initial lecture in that position was delivered October 3, 1850, in front of the university's first medical class, which consisted of over ninety male students and visiting physicians; women would not be admitted to the program for another two decades.

As described by Horace Davenport in his history of the medical school, the curriculum, as it was established in September 1850, consisted of four daily lectures during weekdays and clinical demonstrations on Saturdays, with courses lasting from October to April. To be admitted to the school in that era, an applicant "had to present evidence of a good moral character and sufficient knowledge of Greek and Latin to understand the technical language of medicine."[11] To graduate, the student had to attend two lecture courses and produce a thesis.

The requirement for proficiency in Greek or Latin has long been eliminated for admission to the medical school of the twenty-first century. Nevertheless, a basic knowledge of the Greek or Latin root of medical terminology remains helpful in understanding (and even learning) both modern spelling and the seemingly ever present "phth" diphthong one often encounters; examples include diphtheria or phthisis, the older term for tuberculosis. Each originated from the *phi theta* spelling in the original Greek.

In addition to the positive "moral character," the applicant had to have apprenticed for three years either prior to, or concurrently, with a physician, one not associated with the practice of homeopathy. As Davenport pointed out, the apprenticeship could involve legitimately learning medical procedures, or just as possible, merely "hitching the physician's horse to his buggy, compounding prescriptions, and helping hold a patient undergoing an operation without anesthesia,"[12] a not uncommon situation during that era. The standards of the time were "far below that set by French and German schools, which had high entrance requirements, four or six-year curriculums, professors distinguished for their research accomplishments, and access to large hospitals."[13] Enrollment during this period peaked at 525 students in the year immediately following the end of the Civil War, representing nearly half of the total enrollment at the university.

Though the quality of the university medical school gradually improved during the three decades prior to Vaughan being appointed dean, if an American student had aspirations for a strong medical education, he—and in that era it was universally nearly 100 percent male—was forced to travel to Europe where German, and even French, schools could be found that were first rate. The situation in the United States began to change during the 1880s, the modernization largely beginning in Baltimore where philanthropist Johns Hopkins had donated much of his vast fortune of $8 million to the university named for its benefactor. Approximately half the donation, amounting to about $3.5 million, the equivalent of $70 million today, was directed towards the establishment of a hospital, the other half of the donation designated for the university; the purpose of the money included the establishment of a medical school. That university, Johns Hopkins, began operation in 1876, while the hospital opened in 1889; the School of Medicine opened its doors four years later.

To ensure the highest quality medical training, Dr. William Welch was appointed as the first dean of the Johns Hopkins Medical School. Welch had acquired extensive training and experience while studying in Europe in 1876 and 1877, working with physicians as distinguished as Friedrich Daniel von Recklinghausen, the noted pathologist, physiologist Carl Ludwig, and most importantly for the Welch's later contributions in the field of bacteriology, Julius Cohnheim. The medical careers of Vaughan and Welch would merge during the years of the "Great War" (World War I) so closely that the two would become known as the "Gold Dust Twins."

The sorry state of medical education in the United States as late as the 1880s was aptly described in an address to the Medical and Chirurgical Faculty of Maryland by Canadian physician Dr. William Osler at the society's annual dinner on April 23, 1889. Osler was one of the pioneers in the medical program at Johns Hopkins as it developed during this time, and served as its first physician-in-chief. In his address, titled "The License to Practice," Osler excoriated his profession. The context of his talk should first be noted. The State of Maryland had already rejected the notion of establishing a board of examiners in policing the profession; those medical schools which did exist in Maryland at the time were primarily "for profit" institutions run by the professors and requiring little in the way of actual medical training.

In his presentation, Osler stated, "It makes one's blood boil to think that there are sent out year after year scores of men, called doctors, who have never attended a case of labour, and who are utterly ignorant of the ordinary every-day diseases which they may be called upon to treat; men who may never have seen the inside of a hospital ward and who would not know Scarpa's space from the sole of a foot. Yet, gentlemen, this is the disgraceful condition which some school-men have the audacity to ask you to perpetrate; to continue to entrust interests so sacred to hands so unworthy. Is it to be wondered, considering this shocking laxity, that there is a widespread distrust in the public of professional education, and that quacks, charlatans, and imposters possess the land?"[14]

As a result, medical training in Baltimore, and indeed among a few schools throughout the country, underwent significant changes in the years ahead. The American Medical Association was instrumental in establishing needed standards for a medical education which could match that available in Europe. Dr. Nathan Davis was considered the driving force behind the association, as it was his 1845 resolution before the New York state medical association which called for the national meeting two years later, the beginning of the AMA. Within the resolution introduced by Davis at the 1847 meeting was the following: "That it is desirable that a uniform and elevated standard of requirements for the degree of M.D. should be adopted by the medical schools of the United States."[15]

While there was nothing controversial in the original resolution, political infighting as well as the war which began a little over a decade later delayed implementation of specific requirements for a medical degree. In 1867, in concordance with the AMA meeting then being held in Cincinnati, Davis again introduced a resolution with a more specific listing of standards and requirements. Among these specifics was

> that every medical student be required to study four full years, including three regular annual courses of medical college instruction, before being admitted to an examination for the degree of doctor of medicine; That the minimum duration of a regular annual lecture term, or course of medical college instruction, shall be six calendar months; That every medical college shall embrace in its curriculum the following branches, to be taught by not less than nine professors, namely: descriptive anatomy, including dissections; in organic chemistry, *mate-*

ria medica, organic chemistry and toxicology; general pathology, therapeutics, pathological anatomy and public hygiene; surgical anatomy and operations of surgery; medical jurisprudence and medical ethics; practice of medicine, practice of surgery, obstetrics, and diseases of women and children; clinical medicine and clinical surgery."[16]

At the completion of each series the student would be required to pass an examination; students would also be monitored for attendance in required classes. Among the chairmen of the AMA's Committee on Medical Education during these years was Dr. Alonzo Palmer, a professor of internal medicine and pathology at the University of Michigan Medical School, and twice dean of the school during between 1875 and 1887.

The list of requirements for completion of the medical degree became mired in arguments centered on how realistic it would be to implement such changes, regardless of their desirability; in the time honored practice, the resolutions were subsequently referred to subcommittees for discussion in future meetings. Meanwhile the number of medical schools continued to increase; in 1870 fifty medical colleges were in operation in the United States, a number which doubled by the following decade. By 1890, one hundred and thirty-three medical colleges existed, many motivated for the simple purpose of profit. Still, a few such schools were motivated to implement the standards recommended by the AMA, including Johns Hopkins, as described above, schools in California, and the University of Michigan.

While the medical school in Ann Arbor had not descended to the depth to which Osler had described for Maryland, it too instituted significant improvements during the period between 1850 and 1889, over time keeping it more in line with the changes suggested by the AMA resolutions and establishing the school in the upper tier of such medical programs, at least in the United States.

The first person chosen by the faculty to be dean (at the time the position was referred to as president) of the medical school at Michigan was Dr. Abram Sager (1810–1877). Sager had joined the faculty in an unpaid position in 1842, receiving a more formal appointment several years later; with the establishment of the medical school in 1848 and the beginning of formal instruction two years later, Sager was named professor of obstetrics in addition to his continuing role as dean. Between 1850 and 1875, by which time he retired, Sager periodically served in a position which seemed to most often rotate among the medical faculty rather than providing stability in leadership. Sager was succeeded (albeit briefly) by Dr. Samuel Denton (1803–1860), a physician who also had an appointment as professor of physics, teaching pathology and the theory and practice of medicine as well. Denton was elected dean twice, serving from 1851 to 1853 and 1857 to 1858. His tenure in this position was interrupted by Dr. Silas Douglas, who served as dean between 1852 and 1857, and again from 1862 to 1868. Douglas had been trained in chemistry, arriving in 1844 as an unpaid assistant; his appointment was subsequently upgraded to that of professor of chemistry, mineralogy and geology. Douglas had played a significant role, along with Sager and Pitcher, in persuading the regents to formally establish the medical school. His position at the university came to an untimely end when in 1877 he was accused of embezzling funds paid by the students. Though the Michigan Supreme Court subsequently supported his claim of innocence, he never returned to the university.

Dr. Moses Gunn (1822–1887) served briefly as dean between the years 1858–1859. Gunn had been appointed professor of anatomy and surgery in 1849, bringing with him to the university a cadaver he had acquired while previously teaching at Geneva Medical College in New York; Geneva continues to the present day as a member of the State University of New York (SUNY) system. While the use of cadavers is now commonplace in modern medical schools, at the time of Gunn's appointment it was highly unusual to have any such learning tool available for students; grave robbing was a not uncommon profession during this period. Gunn reportedly placed the body in a trunk, transporting it by stagecoach first to Detroit and subsequently to Ann Arbor.[17] In 1854 he became professor of surgery in the medical school. During the Civil War he served eleven months with the Army of the Potomac under the command of General George McClellan before returning to the university. Following the death of his son in 1867, Gunn resigned from Michigan and moved to Chicago where he was appointed chair of surgery at Rush Medical College. He remained there until his death.

Dr. Corydon Ford (1813–1894) was elected to the position three times, serving in 1861, and then again from 1879 to 1880 and 1887 to 1891. He had received his medical degree from the medical school in Geneva, New York, in 1842, serving as an anatomy instructor there prior to joining the University of Michigan as professor and chair of anatomy in 1854. Ford's relationship with Moses Gunn had a tinge of destiny associated. For a time they had shared a room while Ford was a student at Geneva Medical College. Gunn had predicted at the time that sometime in the future he (Gunn) would be a professor of surgery in a medical school and that Ford would be a professor of anatomy in the same school. With the appointment of Ford at the university, Gunn's prediction had come true.

It was in 1869 that the medical school established a university-owned hospital, the first school in the United States to do so. The building had originally been one of four structures built in 1839 and 1840 as housing for professors at the university. A number of faculty had occupied the houses in the decades after their construction, though the specific occupants of what would become the hospital are unknown. The proposal for conversion of the house was submitted by Cyrus Stockwell, a member of the board of regents during the June 1869 meeting: "The Committee on the Medical Department, to whom was referred the project of organizing a hospital in connection with the Medical Department of the university, would respectfully recommend that the northeast dwelling house on the university grounds be set apart for that purpose."[18] The report was adopted. The facility lacked wards or operating rooms and served primarily as a place for patients to stay prior to examinations by students. It continued to serve as the university hospital until 1891 when it was replaced by a more modern structure on what was then Catherine Street in Ann Arbor. The structure was demolished in 1908 when the new chemistry building was built on the site.

The following year, 1870, the first women were accepted into the medical program, one of only approximately five such programs in the nation at the

Physiological chemistry laboratory (Bentley Historical Library, University of Michigan [image #002053]).

time. "I well remember the day we read in the Boston papers that the University of Michigan had opened its doors to women in all departments," physician Eliza Maria Mosher recalled years later. The year that would forever stand out in her memory was 1870. She and four other interns—Amanda Sanford, Anna Hutchinson Searing, Elizabeth Hait Gerow and Emma Louisa Call, all of whom wanted the kind of medical education available only at the best medical schools—were in the laboratory at the New England Hospital for Women and Children in Boston (established in 1862 for the purpose of providing medical experience for women) when they heard the news. "We five young women joined hands and danced around the table," Mosher said. The women had no illusions about how they would be received at Michigan. Mosher had already applied to Michigan and had received a reply from Professor Alonzo Palmer stating, "For my part, I cannot see how right-minded women can wish to study medicine with men. But the women knew that success in their future careers depended upon receiving a thoroughly competent medical education at an established school and they were ready to take on whatever challenges faced them."[19] There were initial concerns that some of the subjects might offend the "sensibilities" of the female students. To avoid such problems, courses in anatomy and gynecology were taught in segregated sessions to only women; presumably faculty would not object since they were paid an additional $500 for doing so. Whether because the faculty decided such separations were unnecessary, or the university decided the extra cost involved was not worth the trouble, within a year all classes integrated men and women; women still were seated separately from their male counterparts.[20]

All of the women successfully completed their medical education at Michigan and went on to establish outstanding careers in the medical field. On the whole they felt the faculty treated them fairly while they pursued an education. As Emma Louisa Call later remembered,

> The first class of women ... were naturally the objects of much attention critical or otherwise (especially critical) so that in many ways it was quite an ordeal. I believe that only one of the medical faculty [Ford?] was even moderately in favor of the admission of women, so that it speaks well for their conscientiousness when I say (with possibly one exception) we felt that we had a square deal from them all. Corydon Ford had taught Elizabeth Blackwell [the first woman to earn a medical degree in the United States] at Geneva Medical College and was tolerant of the women students, but not all the professors were so kind. Silas Douglas did not intervene when the men students stamped their feet and shouted as the women entered the lecture room for the one subject that was taught to mixed classes.[21]

In 1871 Amanda Sanford [Hickey] (1838–1894) would be the first woman to receive a medical degree from the university, graduating with honors. Her

Deans of Medical School (foreground, left to right) Abram Sager, Alonzo Palmer, Corydon Ford, Moses Gunn and Silas Douglas (Bentley Historical Library, University of Michigan [image #bl002094]).

graduation thesis, "Puerperal Eclampsia," contained a review as well as original research into a serious complication of pregnancy with a significant mortality rate for both the mother and child. Sanford subsequently established a practice in Auburn, New York, as the first female physician in the city, later participating in the purchase of land on which the site of the Auburn City Hospital would be built. Her expertise in the medical field resulted in her acceptance into the previously all male Cayuga County Medical Society, and subsequent election to the presidency of the organization.

Eliza Mosher (1846–1928) graduated with a medical degree from the university in 1875, returning to Poughkeepsie, New York. In the 1880s she returned to Michigan as professor of hygiene and women's dean, where she remained until 1902; the first large women's dormitory on the campus was named in her honor. Anna Hutchinson Searing (1830–1912) graduated in 1872. She continued her education in Vienna for one year. After returning to the United States, Searing established a medical practice in Rochester, New York. Elizabeth Hait Gerow (1845–1933) graduated in 1875, following which she established a joint practice with Mosher in Poughkeepsie. Emma Louis Call (1847–1937) graduated with top honors in 1873, following which she also traveled to Vienna to continue her education. She returned to the New England Hospital for Women and Children in 1875 and practiced there until 1920. Mosher was elected to the Massachusetts State Medical Society, the first woman so honored.[22]

Alonzo Palmer (Bentley Historical Library, University of Michigan [Alonzo Palmer file]).

Arguably the most prestigious member of the university faculty elected as dean prior to the appointment of Victor Vaughan was Professor Alonzo Palmer (1815–1887). Palmer served from 1875 to 1879 and again from 1880 to 1887. Palmer had received his medical degree in 1839 from the College of Physicians and Surgeons, now part of Columbia University, in New York. Prior to joining the Michigan faculty in 1854 as professor of anatomy, Palmer

Chemistry laboratory (Bentley Historical Library, University of Michigan [image #002006]).

had practiced in the town of Tecumseh, Michigan, as well as in Chicago; while in Chicago he had headed the cholera hospital during the outbreak of that disease in the early 1850s. He was editor of the relatively short-lived *Peninsular Journal of Medicine* during the 1850s, in addition to serving as president of both the Michigan Medical Association and the American Medical Association prior to appointment as dean of the medical school. As noted earlier, Palmer had also served as chair of the AMA's Committee on Medical Education which had attempted to establish standards for medical schools. Palmer died "in harness" as it were, in December 1887. He had been appointed by the AMA as president of the Section of Pathology in the International Medical Congress held in Washington, D.C., during the summer of 1887, a period of intense heat during the pre-air conditioning days in that city. Overheated and fatigued prior to the conference, Palmer attempted to recover while vacationing in New Jersey. Returning to Ann Arbor in September, he had still not completely recovered. That December he became critically ill— he had been suffering from a severe bladder infection—after completing his usual lecture before the students, and died just prior to Christmas.[23]

1. Origin and Evolution of the Medical School at the University

Medical school (Bentley Historical Library, University of Michigan [image #002028]).

Corydon Ford succeeded Palmer, the second time he was appointed to that position. Serving from 1887 to 1891, Ford was the last dean elected by the faculty.

By the time of Ford's second appointment the requirements for the medical school had been significantly strengthened. New requirements included many of the changes recommended by Palmer in his role as chair of the aforementioned education committee of the AMA. Until October 1877 the medical curriculum at Michigan consisted of a two-year program. This was changed at that time to a three-year program, although students already enrolled were "grandfathered in" and could elect to continue with the two-year curriculum. The two-year curriculum was officially eliminated after 1880.

The three-year curriculum was codified in 1887. The first year program consisted of ninety lectures in descriptive anatomy, eighty in physiology, and lesser numbers in courses such as chemistry (both in inorganic and organic), histology, bacteriology and embryology. Course work included extensive laboratories as well. Subjects during the second year continued with anatomy (ninety hours), physiology, pathology, the "practice of medicine," obstetrics, surgery and diseases of women and children among requirements. Advanced courses were also available. The third year continued with the "practice of medicine," several courses in surgery, clinical medicine (148 lectures), obstetrics and opportunities in advanced coursework.[24] Other changes were instituted which began with the 1889–1890 class: "All students entering after July 1st, 1890, will be required to spend four years in professional study, including the time spent in attendance upon lectures, before presenting themselves as candidates for the degree of doctor of medicine."[25] The university was the first in the United States to require a four-year program in medicine; other schools followed Michigan's lead and required a similar curriculum in subsequent years.

Additional changes took place that year. The election of Corydon Ford as dean of the medical school represented the last time such appointments came about through the vote of the faculty. In June 1891, Ford, then approaching eighty years of age, resigned, and the board of regents, not the faculty peers, appointed Victor Vaughan as the new dean.[26]

Chapter 2

Victor Clarence Vaughan: The Early Years

Professor Victor Vaughan, the eighth dean of the medical school, was born October 27, 1851, in Mount Airy, Missouri, and died November 21, 1929, in Richmond, Virginia. In his autobiography, published in 1926 some years after his retirement[1] (and a source for much of the biographical information provided here), Vaughan provided an extensive family genealogy as related to him, delving back some nine hundred years. His mother's family was descended from a line of French Huguenots; several early relatives participated in the Crusades, though the family history during these generations is more hearsay than anything based upon documentation. More accurate information begins about the year 1650 with one Bartholomew Du Puy who, as the century ended, immigrated to America, settling in the region of Virginia north of Richmond. It was in this region that the maternal side of Vaughan's family remained and prospered; some were slaveholders. Among these family members was William Dameron, Victor Vaughan's grandfather. In 1829, Dameron migrated west to Missouri with his family; one of his daughters, Adeline Dameron, became Victor Vaughan's mother.

Much less was known concerning the paternal side of the family beyond that of its Welsh ancestry. Vaughan's grandfather, one Sampson Vaughan, arrived in America shortly after the Napoleonic Wars (ca. 1815), purchasing a farm near Durham, North Carolina, where, as a slaveholder, he grew tobacco. Victor Vaughan's father, John Vaughan, rather than simply living on the farm, enlisted in the army, and taking advantage of the "opportunity" spent some years seeing the West.[2] Subsequent to his discharge he settled in Missouri where he married Adeline Dameron. Victor was born October 27, 1851, in his grandmother's house just east of the town of Mount Airy.

As a young boy Vaughan became aware of some of the ravages associated with the Civil War. Missouri was a border state, and though never officially seceding from the Union, it encountered some of the most vicious fighting associated with guerrilla warfare. Several of the most infamous Confederate raiders—"Bloody" Bill Anderson and William Quantrill—carried out some of their actions in that region of the state; Quantrill (spelled Quantrell in Vaughan's memoirs) had dined—uninvited—at the Vaughan dinner table the day prior to that in which his troops massacred captured, and unarmed, Union soldiers who had surrendered. Despite several close encounters with Union soldiers, the Vaughans' losses during the war consisted primarily of horses and household items rather than members of the immediate family. As a result, despite serving in the medical corps in two wars, Vaughan had a lifelong aversion to war or any other form of military conflict.

Vaughan's earliest education, typical of that among boys living on what was still largely frontier, consisted of some formal schoolhouse learning supplemented with what today would be considered home schooling. His first teacher was a local physician, one Dr. William Watts, who lived approximately a mile away. When the Watts family moved away, the locals built a more formal schoolhouse with the fancy name of Hazel Hill Academy; it was here that Vaughan received his earliest formal education. Supplementing this was the opportunity for learning he received at home. Vaughan's mother provided extensive reading material in which a boy who loved to read and learn could easily indulge.

When Vaughan was sixteen, he enrolled in Central College, a Methodist school located in Fayette, Missouri, and now known as Central Methodist University. Conceding his immaturity, Vaughan withdrew after attending for one semester; forty-three years later the by then well-known Dr. Victor Vaughan was awarded an honorary LL.D. degree from the college.

A year later, Vaughan enrolled at Mount Pleasant College in nearby Huntsville, Missouri. At the time a Baptist school, the college was then led by a former Confederate soldier, James Terrill, who ultimately served as both president and owner. The knowledge Vaughan gained from Terrill was primarily Latin, and even that was limited in scope given the shortcomings of Terrill's knowledge in that subject; Vaughan in fact shortly became an instructor himself in that subject.

The greater impact on Vaughan's future while at Mount Pleasant was in the subject of chemistry. Vaughan had stumbled upon a locked room located in the college's main building, and upon obtaining permission to open the room discovered a treasure-trove of chemical supplies and equipment. Likely the materials had been there since before the beginning of the war some dozen years earlier. Using several editions of early chemistry books as texts,

Vaughan taught himself enough chemistry that he was able to teach a course on the subject, establishing his own laboratory as well. Vaughan graduated in 1872 but continued to teach Latin and chemistry for another two years.

Mount Pleasant College survived for only another eight years. A dispute over salary led to the resignation of several teachers. In 1882 a fire, probably the result of arson, burned the main structures and shortly afterwards the college ceased operations.

After a brief sojourn as a teacher at Hardin College in Mexico, Missouri, a Baptist school established for the education of young women, Vaughan was forced to look beyond Missouri for a college or university at which he could continue his education. In the years after the war, a ban was established which allowed only those instructors who had not been Confederate sympathizers to teach at most of the state's public schools. Since Vaughan's family could not meet that requirement, he decided to investigate other schools in which he might enroll or find employment. Among these was the University of Michigan. He enrolled in 1874.

In his autobiography, Vaughan indicated that while he was well aware of the highly regarded reputation of the school, the decision to attend Michigan was based upon the presence of an excellent chemistry laboratory. Vaughan's chemistry education at Mount Pleasant College also owed a significant debt to a particular source and textbook he had obtained there: a copy of *Qualitative Analysis*, first published in 1864 (with newer editions in subsequent years) and authored by Drs. Albert Prescott and Silas Douglas, both chemistry professors at the University of Michigan at the time. In 1874 Prescott was professor of organic chemistry, shortly to be appointed as dean of the college of pharmacy. The potential to interact with such distinguished faculty played no small part in Vaughan's decision.

The chemistry building in which he would continue his studies had been erected during the mid–1850s. By the time of Vaughan's arrival the building had undergone several significant expansions as a result of the prevailing philosophy that all chemistry instruction at the university should be carried out in a single building; such centralization would prove impractical, as science programs at the university underwent significant growth during the early years of Vaughan's tenure. What was lacking in space did not reflect on the quality of the curriculum. The chemistry program at the university was then considered among the best in the world, on par even with those in Germany or Austria.

Vaughan's first challenge was that of simply being accepted into the graduate program at Michigan. He had a bachelor's degree from Mount Pleasant, one considered insufficient for admission to the graduate program in Ann

Arbor. Vaughan was first required to demonstrate competency in chemistry, his desired major, as well as in geology and biology. Prescott himself was chosen to conduct the chemistry interview, which Vaughan passed easily. The second member of the faculty to interview Vaughan was Professor Mark Harrington, who spoke with Vaughan about biology; the interview did not go well, with Harrington reportedly telling Vaughan, "I suppose you know as much about biology as our freshmen do."[3] Professor Eugene Hilgard, briefly serving as professor of mineralogy at Michigan, and considered among the founders of what is now called soil science, was the third faculty member to quiz Vaughan; Vaughan's interview went well. Vaughan became Hilgard's assistant the following year, impressing the professor enough that when Hilgard accepted an appointment in 1875 at the University of California, Berkeley, he wished Vaughan to accompany him; Vaughan declined the offer. Vaughan graduated with a master of science degree in June 1875, with his thesis entitled "The Separation of Arsenic and Antimony."[4]

After declining Hilgard's invitation, Vaughan continued with his graduate work in the newly established doctoral program, graduating in 1876 with one thesis entitled "The Osteology and Myology of the Domestic Fowl," and a second thesis produced from work involved in the separation of arsenic from other metals.[5] That year he was admitted to the program in medicine.

Prior to his graduation and while still working as a teaching assistant in chemistry, Vaughan was appointed to teach a course in physiological chemistry. The appointment was fortuitous, the result of a scandal and controversy which took place within the chemistry program. Professor Silas Douglas of the aforementioned text on qualitative analysis had been instrumental in the original effort to fund a laboratory building devoted completely to teaching chemistry; his efforts resulted in the laboratory building which had first been erected in 1855 during one of his tenures as dean. Douglas continued in his role as chemistry instructor (as well as being elected mayor of Ann Arbor in 1871) until 1878. Dr. Preston Rose taught toxicology and urinalysis to the medical students during this period as well, being appointed assistant professor of physiological chemistry in 1875.

According to the bylaws of the university, "Each student should be furnished with apparatus and chemicals at their cost price, or according to the price list of a New York dealer, and only such chemicals as shall actually be used shall be charged, and the amount thus received by the professor of chemistry shall constitute a fund in his hands for the purchase of apparatus and chemicals for laboratory use, which amount shall be properly accounted for at the close of the year."[6] In other words, students paid a fee, generally about ten dollars, for use of reagents and equipment, the funds for which went to the university. (Despite inflation, the chemistry fee paid by the author nearly a century later was still ten dollars.) It was discovered that laboratory receipts from 1866 to October 1875, a total of $831.10, were unaccounted for (though the confusion actually involved several thousand dollars).[7] Rose was accused of having stolen the money. According to Vaughan, Rose would turn any

Vaughan's description of his job, 1876 (Bentley Historical Library, University of Michigan [Victor Vaughan file]).

money he collected over to Douglas, who then provided an informal receipt consisting of the initial D on a stub, a total consisting of several thousand dollars by the time the discrepancy was alleged.[8] The primary cause of the confusion lay in the informal accounting methods used to keep track of student payments—card vouchers recorded by Douglas which were turned over to the regents at the end of each year; no trained accountant was involved. Rose eventually paid a portion of the amount, while mortgaging his house for the remainder. Though the State Supreme Court, after six years of litigation, eventually placed the blame on both parties for their informal accounting methods, Rose was suspended from the university in December 1875. Vaughan was appointed in his place to teach the course in physiological chemistry, an appointment which was made permanent the following June 1876.

Vaughan's teaching duties as a lecturer during this period of his academic career largely mirrored that of other lecturers in the program. The total number of students in the program was approximately 200; students were divided into three sections. Vaughan's typical daily routine consisted of several hours of lecture to various groups of students, coupled with laboratory instruction in microscopy, toxicology—Professor Rose's specialty—and physiological chemistry. Class times included Thursday evenings and Saturday afternoons. The rest of his time was devoted to the "necessities of life, as eating, sleeping etc."

Chapter 3

Member of the Faculty

Advancement came quickly. In June 1877, while he was still working on his medical degree, "The Faculty of the Department of Medicine and Surgery hereby recommend the appointment of Victor C. Vaughan, Ph.D., to the position of instructor in medicinal chemistry and lecturer on physiology, with an addition to his present salary of seven hundred and fifty dollars."[1] Vaughan's duties included teaching histology and physiological chemistry. There was some reluctance on the part of the board with Vaughan's appointment, with the charge of "atheism" applied to Vaughan's religious beliefs. Alonzo Palmer, dean of the Medical School, informed Vaughan of the charges, to which Vaughan replied that he was teaching science, not religion. Supported by the faculty of the medical school, the charge was ultimately ignored and Vaughan's appointment was approved. That year an additional change took place in the life of Victor Vaughan as well: his marriage to Dora Taylor. The family would someday include five sons.

In March 1878, Vaughan was granted his medical degree, one of sixty-five such students, including six women. He continued teaching physiological chemistry, and received an additional appointment as lecturer on medical chemistry in June 1879.[2] A

Victor Vaughan, 1877 (Bentley Historical Library, University of Michigan [Victor Vaughan file]).

Victor Vaughan's letter of acceptance for a June 1879 offer of position as lecturer in medical chemistry (Bentley Historical Library, University of Michigan [Victor Vaughan file]).

year later he became assistant professor of medical chemistry, and a full professor of physiological and pathological chemistry as well as associate professor of therapeutics and *materia medica* in June 1883.[3]

Establishment of the Hygienic Laboratory

The year 1883 also marked the beginning of Vaughan's evolution in the area of public health. The Michigan State Board of Health had been established in 1873 by an earlier act of the legislature. Dr. Henry Baker was the primary founder and its first secretary. While Michigan was not the first state to establish such a board—that honor went to Louisiana (1855), followed after the war by Massachusetts (1869)—Baker was guided by the precedent of Massachusetts in setting the guidelines for the agency. Vaughan was appointed to the board in 1883 by Governor Josiah Begole and served in that capacity for over thirty-five years. His training and interest in laboratory work became the driving force in Vaughan's emphasis to the board of the need to establish a laboratory.

Vaughan first brought the concept of a hygienic laboratory before the board at a meeting held in Lansing on January 8, 1884. The context was in the question brought by Dr. J.H. Kellogg to the board of health for "using a part of its appropriations for making special investigations" of designated unsanitary conditions [as at the State Reform School]. Since no laboratory yet existed for carrying out such an investigation, Vaughan "spoke of the need for a fully equipped sanitary laboratory at the university." The board subsequently passed a resolution designating that "a sum not exceeding three hundred dollars be appropriated to pay for results of original investigations in sanitary subjects."[4] Despite Vaughan's request for the board to support the creation of a sanitary laboratory, preferably to be built at the university, the board of health as yet took no action.

Among the first public health issues brought to the attention of Vaughan and other members of the board of health was an outbreak of illness seemingly associated with milk and cheese products. Between August 1883 and August 1884, outbreaks of violent illness were reported in various parts of Michigan. That the cause of the illness was cheese was ascertained by determining who within individual households became ill, and what they had eaten; even within families, only those who had eaten cheese became ill. The first report came from Oxford, Michigan, in August 1883, where fourteen or fifteen persons were reported ill. By the time the outbreak ended a year later over 212 persons were reported to have suffered illness. The source of the outbreak was ultimately traced to one cheese manufacturer, the Old Original Fairfield Factory of Fruitridge, Michigan.[5] Samples from the contaminated cheeses were sent to Vaughan's laboratory at the university for analysis. While there was obvious evidence for the presence of toxins or poisons as the specific cause of the illnesses, extensive bacterial contamination of the cheese was also observed.

A more complete analysis of the cause of the outbreak was provided by Vaughan the following year.[6] In his report Vaughan described the isolation and testing of a contaminant observed in samples of cheese. The substance was extracted in alcohol and allowed to dry, resulting in a "fatty mass" which Vaughan carefully tasted [carried out at a time in which it was not unusual for the scientist to serve as his own "guinea pig"]. Within a few minutes Vaughan noted a dryness of mouth and constriction of his throat. Further purification of the substance yielded "needle-shaped crystal" which, when placed on the tongue, produced the symptoms of nausea described by those who had eaten the contaminated cheese. Vaughan termed the crystals tyrotoxicon.[7] Vaughan believed the presence of the toxin resulted from the practices carried out by the farmer in the collection and storage of the milk. "Evidently, tyrotoxicon may originate in milk on long standing in closed vessels. As the putrefactive changes in the milk are due to the growth of minute organisms [observed by Vaughan upon microscopic analysis], the introduction of these organisms into the milk may hasten its putrefaction, and, consequently, the formation of the ptomaine [poison].[8] The germs may be present in portions of milk which adhere to the sides of vessels which are not cleansed as often or as thoroughly as they should be. I would suggest that cheese manufacturers thoroughly inspect the cans in which milk is brought to them. When cows are kept in filthy stalls, the milk is likely to undergo speedy putrefaction."[9]

Working with an associate, Frederick G. Novy (spelled Novie in the cited publication), Vaughan determined that the substance tyrotoxicon was identical to a salt of the toxic material diazobenzol, likely the result of bacterial putrefaction. When a small amount of the purified substance was fed to a cat, the animal developed symptoms identical to those exhibited by victims of the outbreak. The presence of the substance in contaminated foods was also observed in outbreaks of illness following the eating of disparate foods such as oysters and ice cream. Vaughan attempted to identify the microorganism, but could only determine that it probably was some type of an anaerobe.

> We are conducting some experiments with the hope of ascertaining the nature of the microorganism which produces this poison, but are not ready yet to make any definite report on this point. We will only say that it seems to be a germ which develops best in the absence of air or with only a limited supply of air. We have inoculated two samples of milk with the same material, leaving one sample open to the air and placing the other in a stoppered bottle, and after a few days found the poison in the stoppered bottle but not in the open beaker, though this may be explained by supposing that decomposition goes on more readily in the open jar and the poison does not accumulate. Fresh oysters, inoculated with poisonous material and left in an open beaker, became poisonous, and the same was true of some thick custard. Though here, again, the upper layers of the oysters or custard may have prevented free access of air to the underlying portions.[10]

Vaughan and Novy were not able to determine precisely which bacterium might have been the source of the toxin, but it may have been a nitrogen compound by-product produced during butyric acid fermentation, a process carried out as we now know by several species of anaerobic bacteria. Vaughan was in all likelihood correct in his conclusion that the putrefaction resulted from an anaerobic process.[11] As an outcome of the investigation, recommendations were made and subsequently implemented that proper sanitation measures were to be instituted: proper cleaning of hands among handlers, better maintenance and cleaning of stalls, receptacles were to be sterilized and the milk itself must be kept cold when stored or transported. Outbreaks of milk-borne illnesses subsequently became less common.

Novy began a long-time collaboration with Vaughan during this period. He was in his early twenties when he graduated from the University of Michigan in 1886. The following year he was offered a position as instructor in the department of organic chemistry; his collaboration with Vaughan began a lifelong study in the fields of bacteriology and parasitology. In addition to his research, Novy's association with the University of Michigan included that of a future dean of the medical school (1933–1935) as well as chair of the new Department of Bacteriology (1902–1935).

The period of medical history between the 1880s and the new century has come to be known as the "Golden Age of Microbiology." Disease was increasingly shown to be the result of infection by biological agents; techniques for studying these organisms were rapidly being developed, albeit primarily in Europe rather than in the United States. Vaughan was among the early members of the American medical establishment to acknowledge the role of bacteria as disease agents. Vaughan and Henry Sewall, a professor of physiology at the university, reportedly spent many a Sunday stroll through campus during these years discussing the alleged role of bacteria and disease. Vaughan already had experience in teaching a course in 1881 on this subject, an elective course called "Sanitary Sciences." Topics included areas of bacteriology such as disease germs and "antiseptics and disinfectants." This was followed up during the mid–1880s with a series of such lectures in collaboration with weekly talks by Sewall.[12]

As was the case with many noted American scientists during the latter years of the nineteenth century, Henry Sewall (1855–1936) had spent time studying in Europe, where he was exposed to the importance of laboratory experience as a necessity in teaching. Appointed professor of physiology at the University of Michigan in 1882, Sewall is credited with developing the first laboratory course in that subject in an American medical school. Among Sewall's contributions to science was his demonstration that injection of small amounts of a poisonous chemical, snake venom, into an animal—pigeons—can result in establishing immunity to that substance. Pigeons which had previously been exposed to small quantities of the venom were able to tolerate later injections of larger amounts. At the time (1887), Sewall did not recognize the significance of what he had demonstrated. Several years later Emil

Henry Sewall, professor of physiology (Bentley Historical Library, University of Michigan [Henry Sewall file]).

Pierre Charles Alexandre Louis (Images for History of Medicine/ National Library of Medicine).

Behring and Shibasaburo Kitasato carried out similar experiments using toxins derived from the tetanus and diphtheria bacilli which resulted in discovery of serum antitoxin.[13]

In 1888 Vaughan and Frederick Novy, by then an instructor in organic chemistry, traveled to Europe, ostensibly using vacation time, where they planned to work for a time in the laboratory of Robert Koch. Novy's purpose was to observe modern methods in the study of bacteriology with the hope of acquiring expertise in these newer techniques; Koch and his associates had been instrumental in establishing what was known as "the Germ Theory of Disease," the concept that microorganisms are the etiological agents of disease. In 1882 Koch himself had identified the organism which is the agent of tuberculosis, now known as *Mycobacterium tuberculosis*. There was an element of irony in this for Vaughan. In 1872 he had been diagnosed with pulmonary tuberculosis. Treatment as prescribed by a physician uncle consisted of standing each morning in cold water obtained from a local sulphur spring. Whether because of or, perhaps, in spite of, the treatment, Vaughan recovered.

The decision made by Vaughan and Novy to travel to Europe for training in the nascent field of bacteriology in the laboratory of Robert Koch was hardly unique among American researchers or physicians. During the nineteenth century, a period during which medical training in the United

States could accurately be described as abysmal, persons interested in the medical field and wishing to perfect their training routinely traveled to London or Edinburgh in the United Kingdom, and, increasingly, either to Paris or to the education centers in Germany. Until the 1880s, modern laboratories, or for that matter anything passing for advanced education in medicine, were largely nonexistent in America.[14] As pointed out by

Robert Koch (Images for History of Medicine/National Library of Medicine).

Professor's house, first university hospital (Bentley Historical Library, University of Michigan [image #bl000005]).

Thomas Bonner in his study of the subject, the problem was not a lack of medical schools—between 1834 and 1892 over 150 such schools had been established in the United States. It was that medical training had evolved into what was primarily a business, "turning out active practitioners in the shortest possible time."[15] Obtaining a medical degree was an issue of money, as schools competed for students through the practice of lowering tuition or minimizing the requirements for graduation.

By the mid-nineteenth century, the *École de Médecine* in Paris had developed into a center for training American medical students. Pierre Charles Alexandre Louis had become the most noted of the French physicians in charge of training at the *École* during this period. Louis' use of the "numerical method" in the analysis of efficacy of treatments for disease—for example, bloodletting as a treatment for pneumonia—became a forerunner to modern epidemiology. Other work included Louis' detailed descriptions of tuberculosis and typhoid fever. Among the many American physicians who studied with Louis during this period was Henry Ingersoll Bowditch who, among other works, edited and translated Louis' works on phthsis (tuberculosis) and typhoid fever. Oliver Wendell Holmes, the Boston physician who helped establish the role played by physicians in transmission of puerperal fever among postpartum women (and father of the noted jurist) was among other prominent physicians who studied at the *École*. Louis was perhaps the most renowned of the European physicians during these years; as an instructor in medicine he was known for his emphasis on observation and research, rather than the rote practice typical of American physicians. Bowditch once described Louis as "a careful observer of facts, [who] deduced from these facts laws which regulate disease."[16]

By the 1870s and 1880s German universities had largely surpassed those in England or France as centers for medical research, both in the quality of the teaching facilities and in the quality of research which was carried out. "The German university was well suited to advance scientific knowledge. Its traditions of freedom, its flexible yet firm organization, and its laboratories ... gave her a powerful advantage once inductive science was introduced."[17]

Koch himself represented the epitome of the German medical researcher. As related above, during the early 1880s Koch identified the etiological agents of tuberculosis and cholera; in the mid-1880s Koch offered at the Hygienic Institute in Berlin the first public courses in the new field of bacteriology, enabling practitioners of the subject to obtain firsthand experience in new techniques. One of the first Americans to obtain such training at the institute was William Henry Welch, one of the "Big Four" founders of the research arm of the Johns Hopkins Hospital in Baltimore during these years.[18] So it was no surprise that when Vaughan and Novy wished to obtain training in the most modern methods of bacteriological study, Berlin would be their chosen destination.

Vaughan related in his autobiography that they had been advised to first obtain letters from either the president or from the state department in order to be admitted to the German laboratories. Rather than troubling any officials for such documents, Vaughan and Novy simply went directly to Koch's laboratory, where they found there was no problem in being allowed to participate in the work.[19] While in Berlin, Vaughan and Novy also attended lectures on hygiene, at the same time developing expertise in laboratory practices with Koch's associate Dr. Carl Fraenkel.

A request for construction of a sanitation laboratory was again presented to the board of health in October 1886. This time the work carried out by Vaughan in determining the cause of the poisonings associated with contaminated milk and cheese proved decisive.

> The proposition to maintain such a laboratory at the university has come about because of recent valuable work done in the present imperfect laboratory at the university, by Professor Vaughan, who lectures in sanitary science, in the school of political economy, at the university, and whose original investigations into the nature of the cause of numerous cases of poisoning in this state have resulted in learning, not only the nature of that cause, but probably also of the cause of one of the most important diseases of mankind. Professor Vaughan's important researches are already known and acknowledged throughout the civilized world. It is a mistake, therefore, to suppose that it is an entirely new scheme to establish a laboratory of hygiene at the Michigan University. It is not an untried experiment. It is a proposition to give proper room, opportunity and support to a laboratory which has already made contributions of incalculable value for the promotion of human welfare; and a proposition to provide for better instruction in a subject now imperfectly provided for, but which is the most important of all subjects which receive attention at the state university.[20]

A resolution was then unanimously passed:

> *To the Honorable the Senate and The House of Representatives*
> Your memorialists, the members of the State Board of Health, respectively represent that:
> *Whereas*, the highest education and that of the most use, is that which bests fits mankind for right living, that which tends directly to the preservation of life, and to the perfection of physical and mental health and strength; and,
> *Whereas*, The teaching of knowledge "of most worth" at the University of Michigan is not yet well provided for; therefore,
> *Resolved*, That we earnestly memorialize your honored bodies to take such action as shall lead to the maintenance of a well-equipped laboratory of hygiene at the University of Michigan, and of such instruction in sanitary science at that Institution, as shall place that subject on a plane not inferior to that of any other subject taught at the university.[21]

The initial request by the regents before the legislature was for an appropriation of $75,000 toward laboratories for physics, hygiene and other medical programs. The appropriation, as passed by the legislature, was reduced to $35,000, to be used "for the construction of a building for scientific laboratories and for the equipment of the same for the year eighteen hundred and eighty-seven."[22] While the money appropriated by the legislature was a beginning, realistically it was insufficient for construction of the building necessary to house the desired laboratories. Therefore a committee consisting of Vaughan, Professor A.B. Prescott and Dr. John Langley, professor of chemistry, presented to the board of regents, first a request that a department of hygiene be established at the university, that hygiene be united with physiological chemistry under one head with the title of professor of hygiene and physiological chemistry, the chair to be provided with adequate assistance, and the additional appointment of an instructor in hygiene and physiological chemistry. Vaughan was appointed chair, with the title of professor of hygiene and physiological chemistry and the director of the hygienic laboratory.[23] Dr. Frederick Novy, having recently graduated with his doctorate, received the appointment as instructor in hygiene and physiological chemistry.

The establishment of a director for the laboratory did not resolve one obvious problem: There was no laboratory to house the department, and the $35,000 appropriation from the legislature was not adequate to build one specifically for that department. The dilemma was solved by a compromise. Regent James Shearer, chairman of the Committee on Buildings and Grounds, pointed out that the $35,000 appropriation was not entirely for construction of the building; one-fifth, $7,000, was to be used for purchase of equipment, leaving only $28,000. The committee recommended "that a brick building be erected, three stories in height, and on a plan suited to accommodate at least two of the laboratories, those of hygiene and physics, and that this proposed building be respectable and appropriate in design, and not to exceed in cost the sum of $28,000." The regents increased the sum to $30,000.[24] As a result, the board of regents passed a resolution "that the Committee on Buildings and Grounds be and are hereby authorized to procure and decide on designs and specifications for the two buildings intended for physics and hygiene in the one case and for anatomy in the other."[25]

The firm chosen to provide the plans for the new building was that of Pond and Pond. Established by brothers Irving and Allen Pond, both graduates of the university, the firm was noted for its architectural design of university buildings; the firm also designed the house the Vaughan's lived in beginning in 1882. Construction was completed by the time Vaughan and Novy returned from Europe during the fall of 1888; the laboratory was opened on October 1 and completely ready for occupancy by January 1889. What was known for many years as the West Physics Laboratory housed the Department of Physics in the basement and first floor; the Department of Hygiene was on the second floor and attic. The building remained in use until it was demolished in 1966. The Hygienic Laboratory was the first of its kind established in the United States, and only the second in the world; Professor Max von Pettenkofer, whom Vaughan had the opportunity to meet while in Germany, had established the first in Munich. Vaughan applied much of what he had observed or learned while in Europe in establishing a state of the art (circa 1889) laboratory which contained the latest sets of apparatuses then available; included among the instruments were Zeiss and Leitz microscopes, developed in Germany during the 1840s and containing the Abbe Condenser, a staple in many microscopes even into the twenty-first century. Animal facilities were also constituted among the facilities of the Hygienic Laboratory.

As described earlier, in July 1887 the regents had established a professorship of hygiene. Vaughan had learned the latest bacteriological techniques while visiting Koch's laboratory, becoming one of the few physicians in the country trained in the nascent science of bacteriology. Consequently he was appointed as the first director. Students who wished to study with Vaughan were selected purely on the basis of wishing to carry out original research, and were expected to complete the course in three months. The program was intense, as students spent a minimum of four hours on a daily basis working in the laboratory. The hygienic course was originally separate from the actual medical curriculum during the first year in which it was instituted; in 1890 the hygienic course was incorporated into the second and third years of the medical program.

The purpose of the laboratory was three-fold: identification of the etiological basis of disease, the examination of food or water for possible contamination, and third, the training of students interested in the study of hygiene. The laboratory was not only the first of its kind in the nation, but until 1907 served as the official state hygienic laboratory.

One of the first medical problems addressed by Vaughan as head of the Hygienic Laboratory was the presence of typhoid fever in Michigan. In 1888 over 1500 cases of typhoid were reported in the state, with over 300 deaths; the following year, 1889, over 2500 cases were reported, with over 400 deaths.[26] While specific numbers varied from year to year, with again with over 2500 cases reported in 1896, the average through this period remained approximately 2000 cases each year. The Hygienic Laboratory's role was to isolate and identify the etiological agent and establish the source of the outbreak. The result was one of the few times Vaughan would seriously err in the interpretation of his findings.

Vaughan reported the results of the inquiry into the basis for the typhoid outbreaks at the 1892 meeting of the Association of American Physicians, an organization founded several years earlier by Canadian physician Sir William Osler. Vaughan had first obtained one hundred and forty-eight samples of water from physicians around the state who had treated victims of typhoid. He then attempted to cultivate and identify—at least the morphology—of microorganisms isolated from these samples. Vaughan divided his water samples into three classes: samples which were alleged to have been associated with the outbreak of typhoid, samples which were not believed to have been the immediate source of the outbreak but which were in unsanitary areas, and samples not thought to be associated with illness. Vaughan observed a wide variety of morphological types, the pathogenic potential of which was confirmed by inoculation into animals. As controls for comparison, Vaughan used cultures of *Eberthella*[27] which he brought with him from the Berlin Hygienic Laboratory and which had been isolated from typhoid victims.

Vaughan observed what he considered significant differences between Eberth's bacillus and the isolates from the water sample. More specifically, his conclusion was that "there are marked morphological and culture differences between Eberth germs from spleens in different epidemics of typhoid fever, and that the most skillful bacteriologists have reported most diversely upon the reaction of the Eberth germ with staining reagents, the evidence becomes sufficient to convince me that the Eberth germ, as found in the spleens and other organs after death, is not a specific microorganism, but is a modified or involution form of any one of a number of related germs."[28]

Vaughan's assertion that typhoid fever originated from an infectious agent, specific or not, was itself representative of the evolution in understanding that particular disease and germ theory in general. Until Koch and others firmly established the concept of "germ theory of disease," the prevailing viewpoint among many physicians was that typhoid resulted from an undefined decomposition in the intestines which had taken place in typhoid patients.[29] The isolation of the typhoid bacillus by Karl Eberth in 1880 from the spleen and mesentery of typhoid victims had established the likely bacterial nature of the disease, though it would be several years before the bacterium would be cultured and characterized in the laboratory.

Vaughan's viewpoint, that a variety of germs may be associated with typhoid, engendered significant disagreement at the meeting. In particular, Dr. William Welch, the aforementioned dean of the medical school at Johns Hopkins, took issue with Vaughan's interpretation.

> Dr. Vaughan's statement that a number of bacteria obtained originally from water appeared to be very unlike the typhoid bacillus ... and that the same bac-

teria, when obtained from the animal after inoculation, grew like the typhoid bacillus, is, if confirmed, an interesting and remarkable observation. There is, however, a possibility of fallacy there which he has probably had in mind. A great many of these bacilli produce intestinal changes, and under these circumstances the *bacillus coli communis* [later termed *Bacterium coli*, and finally its present terminology, *Escherichia coli*, named for its discoverer Theodor Escherich] is very likely to enter the organism. In this case one can readily obtain from the spleen and other organs the bacillus coli communis which may be mistaken for the microorganism injected.

Dr. Vaughan is somewhat more impressed with the extent of variation of the Eberth's bacillus than I have been. It is perfectly true that one may obtain from the spleen of typhoid fever patients a number of different bacilli, and as a rule one can obtain in cases of typhoid fever, not, perhaps, so often from the spleen as the kidney or lung, the *bacillus coli communis*. It is a regular invader in typhoid fever, and there is no doubt in my mind that a number of those who have written about the characters of the typhoid bacillus have been working with the bacillus coli communis and not with Eberth's bacillus.

It is interesting that Dr. Vaughan has found so large a number of waters suspected of causing typhoid fever to contain organisms capable of killing animals when inoculated from beef-tea cultures [used to culture the organisms]. They were mixed cultures, I presume. Unless one adopts the view of Dr. Vaughan, which is at present hardly likely to obtain general acceptance, that typhoid fever is not due to any one specific germ but to a variety of more or less related germs, it is difficult to see what the bearing of these experiments is upon the etiology of typhoid fever.[30]

Welch's interpretation of Vaughan's findings, in other words, was that Vaughan was working with mixed cultures from water samples containing a variety of microorganism, rather than only the Eberth (typhoid) bacillus. It is of course now known that the Eberth bacillus, currently known as the genus and species *Salmonella typhi*, is indeed the etiological agent of the disease.

The difficulty for Vaughan in assigning a specific role for the Eberth bacillus was the result of a problem frequently encountered; the actual numbers of the agent found within the victim could be relatively low, often only a small proportion of the total number of organisms present in the tissue. As recently as 1889, three years prior to the confrontation between Vaughan and Welch, French physicians Rodet and Roux claimed that in the abscesses of a typhoid patient diagnosed with peritonitis, as well as another typhoid patient with liver abscesses, they had isolated the bacillus coli communis, arguing the two organisms were actually variations of the same bacterium. Adding to the confusion was the finding that this same organism could also be the agent of a different cholera-like disease. Not only was the organism isolated from typhoid patients, it had been demonstrated to have actual pathogenic potential.[31]

The distinction between the two organisms was demonstrated several years later by London physician Edward Klein. Klein isolated and grew in pure culture each of the organisms; he further distinguished them on the basis of differences in biochemical characteristics. Klein also demonstrated that the distinct morphology of the organisms did not change even after the organisms had been passaged through animals. Klein observed that *bacillus coli communis* represented a significant portion of the intestinal flora, or at least that portion which was easy to cultivate. He also anticipated the future use of *Escherichia coli* as a surrogate marker for fecal contamination: "For, although without the typhoid germ the bacillus coli, we maintain, is not capable of causing typhoid fever, the presence of bacillus coli in water nevertheless indicates a probable pollution with excremental matters, and amongst them possibly with specific—that is, typhoid—excremental matter."[32] Vaughan himself would later apply this same idea of using the common colon bacillus as a marker for fecal contamination. Vaughan was also incorrect on another account in his dialogue with Welch:

> I do believe, and I think I have good reasons, derived from both laboratory and clinical experience, that typhoid fever may originate without a pre-existing case of typhoid fever. The conditions under which one studies this subject makes all the difference in the world. Most of you see this disease in the city. I see it principally in the country. I have seen cases of typhoid fever scattered among the farming population, among men, women, and children who have not been off their farm for weeks; and it would take a good deal to convince me that there was a common source of infection in these cases. I know that some will say that there has been a walking case of typhoid going through the country and that he has polluted all these wells, but I do not believe it. I believe there are different germs, or different varieties of the same germ, that may cause typhoid fever, and that they are widely distributed.
>
> I am satisfied that there is a tolerance for these germs [which] is very easily acquired, and this has suggested itself to me as a possible explanation of the spread of typhoid where there are apparently no cases.[33]

Vaughan alluded to the possibility of typhoid carriers, individuals infected with the bacillus but who show no sign of the disease. In a sense Vaughan contradicted himself in these last two statements, the first expressing his belief that no common source explained the outbreak among isolated farms, yet in the latter explanation allowing for exactly that possibility. This would subsequently be shown to be correct, that carriers showing no sign of illness may still be the source of outbreaks. The classic example of a carrier was that of the cook Mary Mallon, better known as "Typhoid Mary," several years in the future.

Chapter 4

Dean of Medicine

Vaughan's next career move was his appointment by the board of regents to dean of the School of Medicine. The previous dean, Coryden Ford, had been elected four years earlier by the faculty upon the death of Alonzo Palmer. Ford, by 1891 nearing eighty years of age, resigned in June of that year. In the past the position had been filled through election by the medical faculty; Vaughan's appointment was the first in which the decision was made by the board of regents.

The issue was not without an earlier element of controversy. There was a feeling among some that Ford's resignation was not entirely due to his advanced age and a need for "young blood" on the faculty. There was an increasing advocacy for extending the medical school training to four years, keeping this more in line with training in Europe and reducing the time spent working with an advocate—Ford had opposed such an extension, fearing the result would be a significant reduction in the number of medical students—and the issue of whether homeopathy should be included as a branch of medical training.[1] Whether any of this this played a role in Ford's resignation, or whether he had simply grown too old to devote the time and energy necessary to do justice to the job—Vaughan had already taken over many of the duties as dean—was in the end immaterial. The board felt the increasing size and complexity of the medical school required a more centralized control; Vaughan became the popular choice, and after rejecting an offer for an appointment as professor of hygiene at Bellevue Hospital in New York, accepted the offer.

Vaughan quickly had to deal with a number of challenges, not least of which was the replacement of four faculty in the medical school: Dr. Edward Dunster, a professor of obstetrics, had died several years earlier, while Dr. Henry Sewall, a professor of physiology and close friend of Vaughan's, was diagnosed with tuberculosis and forced to resign; Sewall did subsequently recover but still felt compelled to resign his position; Professors Donald MacLean and George Frothingham were dismissed from the faculty.[2] Within several years Vaughan was also forced to deal with a member of the faculty deemed "incompetent," Dr. Heneage Gibbes, a professor of pathology. Gibbes had originated from the old school of thought with respect to disease, and despite clear evidence by the 1890s that the etiological basis of much of disease, including tuberculosis, was the result of bacterial infection, Gibbes remained convinced otherwise.

Despite acquiring the pejorative label of "incompetent," perhaps unfairly, Gibbes could more likely be considered one of the early free thinkers. Born and raised in England, his youth was spent on board ship, allegedly fighting in the opium wars and combatting pirates, and rising to command his own ship by the age of twenty-one. He was trained in the medical sciences, eventually teaching histology at the Westminster Medical School in London and serving as cholera commissioner to India in 1884. He was the author of numerous medical papers on the subject of histology as well as author of a pathology text. While in Ann Arbor he published some two dozen papers in professional journals. Gibbes was considered an excellent instructor by his students as well.[3] Regardless, in 1895 he was finally forced to retire.

An extensive discussion of the new faculty added or replaced during these years is found in Vaughan's autobiography.[4] Some of the newly recruited or hired faculty were particularly noteworthy and deserving of recognition.

The association of Frederick Novy (1864–1957) with Vaughan during the late 1880s was described earlier. After earning both his bachelor of science and master of science degrees at Michigan, Novy continued his studies while studying with Vaughan, earning his medical degree in 1891. He was appointed assistant professor of hygiene and physiological chemistry following his graduation. As a bacteriologist, Novy was best known for his studies of trypanosomes, the etiological agent of diseases such as sleeping sickness, and studies of spirochetes such as those associated with relapsing fever. However, when a possible outbreak of plague took place in San Francisco in 1900, Novy was asked to join a commission to investigate whether plague actually was present and spreading. The commission did determine plague was present, and Novy was instrumental in developing means to address and contain the disease. This led to an unusual occurrence of a laboratory acquired infection the following year. The sources of the disease, its course and subsequent resolution have been clearly described.

> A case of pneumonic plague occurred in the laboratory of Dr. F.G. Novy at the University of Michigan in 1901. Two medical students, C.B.H. and J.G.C. [James G. Cumming], were studying cultures from San Francisco in animals. C.B.H.

reported to his associate and roommate (J.G.C.) at 2 p.m., April 3, that, since that morning, he had been ill and febrile. He had had a dry cough which became productive of blood-streaked sputum at about 10 p.m. Dr. Novy made a presumptive diagnosis of plague on the basis of sputum examination; it was confirmed later by inoculation into mice.... The route of infection considered to be most likely was through cigarettes which the patient rolled while working. He was sent to an isolation room, and his roommate nursed him there. C.B.H. was severely ill and periodically delirious. Fever continued in a hectic manner for about two weeks.... Cough produced about one-half cup of sputum per day.... The patient had no vaccine but was given a "protective serum" during the second week of disease. J.G.C. finally left isolation after 31 days, and the patient left after 34 days. No secondary cases occurred. Subsequently, the patient developed signs of congestive heart failure. This reportedly limited his activity, although he practiced medicine in California until his death some fifty years later.[5]

Novy was also likely to have been the inspiration for the character of Max Gottlieb in *Arrowsmith* (1925), the novel written by Sinclair Lewis which described an outbreak of plague. By the time he retired in 1935 Novy had been the recipient of numerous awards and honors as well as having served as dean of the University of Michigan Medical School from 1933–1935.

Dr. George Dock was hired in 1891 as a professor of medicine. Dock had received his medical degree in 1884 from the University of Pennsylvania in Philadelphia, where he had the opportunity of studying with William Osler. As with many young physicians interested in new areas of clinical and laboratory research, Dock traveled to Germany and Austria where he worked with, among others, Paul Ehrlich and Rudolf Virchow. Upon returning to the United States, Dock was recruited by Osler for the position of assistant pathologist at the newly established University Hospital in Philadelphia. Osler paid Dock a supreme compliment in describing him as "a man who knows more about clinical procedures than anyone in the United States."[6] After an interim period of three years at the University of Texas, Dock was recruited by Vaughan for the medical school at Michigan in 1891, where he remained until 1908. In 1899, Dock established the position of clinical clerkship which allowed medical students to directly care for patients while supervised by members of the faculty. Students were able to attend to patients admitted to the wards, one of the earliest "hands-on" approaches in learning clinical medicine. A similar clinical clerkship had been developed several years earlier by Osler, by then at Johns Hopkins, and likely part of the inspiration for the program established by Dock. The medical school by this time had already expanded, moving into a newly built University Hospital in 1891.

By the time of Vaughan's appointment the debate over whether women should be accepted into the medical programs had largely been settled, even if reluctantly by some of the older faculty. Between 1870 and 1900, over four

Original medical school (Bentley Historical Library, University of Michigan [image #bl004782]).

hundred women graduated within the various medical departments. Among the most prominent was Alice Hamilton, who received her medical degree at Michigan in 1893. Hamilton completed her internship in Minneapolis, following which she studied in Germany for two years. Returning to the United States in 1897, Hamilton continued her studies at the Johns Hopkins School of Medicine. In 1919 she became the first woman faculty member appointed to Harvard University's medical school.

The other significant changes instituted by Vaughan during the immediate years following his appointment were those dealing with the admissions requirements for the medical school and modernization of the curriculum. As was the situation with most medical schools, the basic minimum requirement was graduation from high school. Vaughan did not change this requirement. However, since the quality of such secondary education could vary between schools, the medical school would accept applicants from only those schools certified by the University Literary Department.

Beginning in 1890, the curriculum for the medical school was expanded to a four-year program.⁷ Since few of the students had training beyond high school, the first-year program dealt primarily with those subjects normally addressed in a freshman year college curriculum: mathematics, general chemistry and physics. Only in the second year were the more advanced medical subjects taught. The changes in the curriculum had an immediate impact on enrollment. In 1891, the final year in which the three-year curriculum was available, over one hundred students graduated. Five years later, graduation was reduced to fifty-two students.⁸ Once the more rigorous program adopted by Michigan became standard among medical schools, enrollment gradually returned to its earlier levels.

Victor Vaughan, 1893 (Bentley Historical Library, University of Michigan [Victor Vaughan file]).

As described previously, outbreaks of typhoid fever arguably represented the most important medical problem taking place in Michigan on an annual basis. By the 1890s it had become clear to most physicians that whatever the etiological agent—and Vaughan himself had remained skeptical for some time as to the role played by Eberth's bacillus—the source of most outbreaks was a contaminated water supply. Even the precise level of prevalence of typhoid was mired in some confusion. Often only fatal outcomes were reported to health departments, causing the actual number of cases to be vastly underestimated. In addition, misdiagnosis was common. Simultaneous infection with the typhoid bacillus and the (unrelated) malarial parasite had first been reported by Alphonse Laveran in 1884, with similar cases diagnosed during the following decade.⁹ The illness was given the name typho-malaria. However, a diagnosis of what had been a rare simultaneous infection was often misrepresented in a significant number of typhoid cases; some years reported more cases of typho-malaria than typhoid itself. It remained for the typhoid commission—established some years later as a result of typhoid outbreaks during the Spanish-American War and consisting of physicians Victor Vaughan, Walter Reed and Edward Shakespeare—to determine that what had often been referred to as typho-malaria was actually a single disease, typhoid fever.

The Chicago World's Fair of 1893, also known as the World's Columbian Exposition, was among the subjects of consultation in which Vaughan took part. Daniel Burnham, one of the chief architects involved in designing the buildings and layout for the exposition, which included the water supply, was particularly concerned about the water to be used by the fairgoers. Burnham had good reasons for concern. The Chicago River, source of much of the city water, was notoriously polluted. In part this was the result of poor, some would say ignorant, planning of the city engineers. In the aftermath of the Chicago Fire of 1871 the flow of the river was reversed; instead of flowing into Lake Michigan, the Chicago River was diverted to the Des Plaines River and ultimately to the Mississippi. The concept seemed to work fine until heavy rains caused the flow to reverse, dumping not only raw sewage but dead animals into the water supply. It was not unusual for intake cribs of city water to turn black with contamination.[10]

Typhoid fever was probably the most common of the water-borne diseases with which Chicago residents, like those in many cities, had to cope on an annual basis. In 1890, over 1,000 deaths were attributed to the disease, an increase of some 150 percent from previous years. The number of deaths doubled again the following year to nearly 2,000. The estimate of the total number actually infected ranged to as high as 20,000 persons.[11] The source of the problem was obvious from a recent observation: "It is beyond question that the sewage of about 70,000 persons is daily poured into the lake from Chicago, Lake View and Hyde Park.

"The sewage of fully 400,000 additional population for at least 40 days in August, September and November [1885] was also emptied into Lake Michigan. Twice during this period the indescribable filth of the South Fork was swept into the same source of water supply, and during a portion of the time the contents of the North Branch were pumped through the Fullerton Avenue conduit into this general receptacle and fountain."[12] The response by city engineers was that of "makeshifts and expedients," in the words of William Sedgwick, then a professor of biology at the Massachusetts Institute of Technology, and Allen Hazen, an expert in hydraulics, in their report on the subject.[13]

Prior to the opening of the exposition, Vaughan was asked to serve as one of the consultants in ensuring a safe and sanitary water supply. As he pointed out, at that time there were thirty public sewers dumping their untreated contents into Lake Michigan, to say nothing about an even larger number of private sewers releasing their contents into the lake. Vaughan and the other consultants correctly decided that, given the level of pollution, it would be impossible to ensure a sanitary water supply using the water at hand.[14]

Burnham's solution was straightforward: the water to be used by the fairgoers had to be either purified on spot, or would have to be pumped from elsewhere. For small numbers of persons, including the workers at the site, Burnham instructed his sanitary engineer, William MacHarg, to build a water sterilization plant right on the grounds. The plant utilized water pumped from the lake, then both aerated and boiled the water prior to use. The sterilized water was then stored in large tanks placed around the park, with fresh water added daily.[15]

The plan worked fine for the small numbers of persons involved in creating the fairgrounds, but was not practical for the hundreds of thousands expected for the exposition. Prior to completion of the building, Burnham arranged for pumping fresh water from the springs of Waukesha, a small town in Wisconsin some one hundred miles away. Needless to say he ran into significant opposition from the townspeople, and was ultimately forced to obtain the water from the springs of Big Bend, a town some twelve miles south of Waukesha. Burnham's idea was a success, and whatever problems may have been associated with the exposition, an outbreak of typhoid was not among them.

Water and food analysis (Bentley Historical Library, University of Michigan [image #002065]).

Between 1889 and 1899, the Hygienic Laboratory analyzed over seven hundred samples of water, primarily from various sites in Michigan or nearby Canada, but also from places as far away as the territory of Arizona.[16] Not all the bacteria which they detected were potentially pathogenic. But a significant number of samples did contain possible toxigenic bacteria. Vaughan often sent by telegraph to those in charge of the supplies the results of his analyses, a practice which caused no end of problems as they quickly became public knowledge. That, along with other financial issues, eventually resulted in the State Board of Health establishing a new state laboratory in Lansing to replace the one in Ann Arbor. The laboratory in Ann Arbor continued for many years as a site for instruction until its destruction by fire in 1967.

Vaughan during this period also began researching the germicidal activity of a substance he called nuclein. The German immunologists Emil Behring and Paul Ehrlich had noted the germicidal characteristics of blood serum, work which would eventually lead to an understanding of the role played by soluble proteins such as antibodies and the components of the complement system.[17] Vaughan as well believed that the bactericidal activity of serum, as demonstrated by German scientists or associates working in their laboratories, was found in the protein components of serum. However, when serum proteins underwent peptic digestion the germicidal activity appeared to remain. Since the substance he called nuclein was still present, Vaughan's conclusion was that it was in this non-protein material that the antibacterial activity he was detecting resided.[18]

Vaughan and his student, Charles T. McClintock, subsequently undertook a series of investigations in which they tested the germicidal activity of this serum component which, unlike antibodies, did not have the properties of protein. Extracts prepared from blood and organs obtained from various animals as well as extracts prepared from yeast were found capable of killing organisms as diverse as the anthrax bacillus and *Staphylococcus pyogenes albus* (*Staphylococcus epidemidis*).[19] When Vaughan tested these nuclein extracts on humans he did so with the belief that any benefits would be the result of increased production of leukocytes; indeed some preliminary experiments showed exactly that. However his colleague George Dock, a pathologist and clinician at the medical school whose expertise in blood work was developed while working under the supervision of Dr. William Osler, found no evidence for any increase in white cell count following such treatment.[20] The question remained whether nuclein could serve as a palliative treatment for illnesses. The results were generally inconclusive when nuclein was tested by Vaughan on humans associated with his medical practice, with the exception that in a few cases which Vaughan reported, inoculation may have proven modestly beneficial. A few examples were cited by Davenport, including a ten-year-old girl with tonsillitis who was cured in one day after gargling with a nuclein solution, and a three-year-old boy with "streptococcus diphtheria" who, when he refused to gargle a solution of nuclein, was cured with a spray solution of the substance.[21]

Nuclein by this time was recognized as a component of the cell nucleus, and consisted of a combination of nucleic acid (now known to be DNA) and protein, a fact acknowledged by Vaughan. The discovery of the actual role played by "nuclein" in heredity would not be made for another half century. For a time the pharmaceutical Parke, Davis and Company in Detroit manufactured and sold a nuclein solution, promoting what they called "nuclein therapy." If one is to believe the testimonials, a compilation of results published by Vaughan in *Medical News* between February 27 and March 27, 1897, nuclein would have appeared to have been the newest wonder drug. Vaughan's descriptions were those of seventy-six cases he treated between May 1893 and December 1895. Over 90 percent of the cases (seventy) were diagnosed as pulmonary tuberculosis. At least nineteen of the seventy reportedly, according to Vaughan, received significant benefit from the nuclein injections. Of the thirty who had died, nine were described as having shown some improvement during the nuclein treatments. Five of these seventy-six cases were diagnosed with urinary tuberculosis, including the following example. Of these, four reportedly benefitted from the therapy.[22]

Dr. A.J. Rosenberry wrote,

> If the theory of phagocytosis be true, the great object to be obtained by the therapeutist is to furnish all the cells of the body with the appropriate pabulum, and to increase the number of white cells. That we have in nuclein a substance which furnishes this cell food in a form in which it can be readily utilized, seems quite certain. That it increases the number of white blood cells enormously has been demonstrated by the work of Vaughan, McClintock and Warthin.
>
> Having had experimental knowledge of the action of nuclein in my own case, I wish to record it, in the hope that others with larger opportunities for testing its merits may be induced to give it a trial, for, living, as I do, in a belt that is practically non-tubercular, opportunities for testing it are not frequent.
>
> On February 28, 1894, I passed a large quantity of blood and some pus from my urine.... Two of my colleagues agreed upon the diagnosis of renal congestion.... Meanwhile a specimen had been sent to Dr. V.C. Vaughan, of Ann Arbor, and tubercle bacilli were detected.... I at once proceeded to Ann Arbor to place myself in his hands for treatment. He proposed the nuclein treatment, of which I had not even heard at that time.... On March 20, 1894, the first injection of ½ drachm was given.... From April 1 to November 20, 1894, daily injections were taken. My general health steadily improved; the local symptoms, which at no time amounted to more than slight uneasiness and a dull, heavy feeling at the neck of the bladder, disappeared; only one hemorrhage occurred. At present my general health is excellent.[23]

At least one member of Vaughan's medical team benefited from this work. In 1894, Parke, Davis and Company hired McClintock to organize a laboratory for the manufacture of diphtheria antitoxin.

Though its use as a germicidal substance has been superseded by other more modern methods, nuclein obtained from wheat germ has reportedly been shown to enhance immune responses through mechanisms which are unclear, leading to suggestions by some that nucleotides might even serve as a food supplement.[24]

Vaughan observed another antimicrobial activity associated with hyperleucocytosis, an abnormally high white blood cell count induced by the injection of yeast nuclein into the bloodstream of a dog. Blood removed from the animal in the hours after inoculation was significantly germicidal when incubated with two different species of bacteria: *Bacterium coli* (now *Escherichia coli*) and *Staphylococcus pyogenes aureus* (now *Staphylococcus aureus*). This in itself would not be surprising; an increase in white blood cell numbers would be expected to result in an increase in antimicrobial activity. However, Vaughan also observed that upon heating the blood samples at 55° C, there was a significant decrease in antimicrobial activity against either of the two species of bacteria.[25] What Vaughan likely was observing as the source of antimicrobial activity was the effect of the serum components comprising the complement system, a series of several dozen proteins which function in a cascade fashion in the inhibition or killing of bacteria. Complement or alexin, the name by which it was originally known, had been discovered about the same time Vaughan was carrying out this work by the scientist Jules Bordet working at the Pasteur Institute in Paris; Bordet also reported the heat sensitivity of the complement components. Since Vaughan made no reference to complement (or alexin) in the publication, it was likely he had been unaware of Bordet's work at the time.

In the years leading to the Spanish-American War (1898), Vaughan spent his time not only establishing a first-rate medical school, but obtaining funds—always in short supply—for teaching in the medical program, research, and developing and maintaining the medical library. The medical library which existed when Vaughan first entered the school during the 1870s barely met the definition of the term. The university library had been started in 1840 when Dr. Asa Gray, briefly a professor of natural history (see Chapter One) at the university prior to leaving for Harvard, purchased $5,000 worth of books while traveling in Europe; it is unknown whether medical books were included in this purchase, but since Gray had some medical training by then it is certainly possible some were included. The first known appropriation for the medical library was in 1854: a total of $66; six years later the medical school subscribed to twenty-four journals. The medical library budget during most of the 1860s amounted to between $100 and $500 annually, enough to maintain subscriptions for at least the most important journals, many of which were European. The budget was significantly reduced during and after the Civil War, which meant purchases were increasingly limited. By the mid-1870s, the period of Vaughan's arrival, the medical library consisted of some 1,500 volumes. Earlier issues of journals, pre–1860, were often unavailable.[26]

Few medical journals were even in existence at this time and many of those were published in either French or German. None emphasized research in the clinical area. The medical library at Michigan by the latter half of the 1870s consisted primarily of textbooks, many of which were out of date. The major reason for the deficiency was hardly unique to that time period: money. Though monetary support had increased since the lean years of the war, each faculty member of the school still had a budget of less than three hundred dollars for purchase of material for his department, with the total budget amounting to less than three thousand dollars.

Even in his role as a laboratory assistant in those early years at the school, Vaughan attempted to correct the deficiency. After suggesting it would be more efficient for a single individual to oversee purchases of books or journals for the library, Vaughan was asked by the faculty to take on that job. After consultations with members of the faculty, Vaughan decided the money could be better used for subscriptions for the latest scientific journals rather than the purchase of textbooks. In his autobiography, Vaughan related how he carried out "the only premeditated fraud he ever perpetrated on the University of Michigan" in making his purchases.[27] The librarian at the time was the Rev. Andrew Ten Brook, who served in that position from 1864 to 1877.

> This dear old man [64 years old at the time] was grouchy, one of that class in whom I have been wont to say the milk of human kindness has undergone the lactic acid fermentation. To him I went with my carefully prepared list of journals. He received me with scant courtesy. I think his mood was partly due to the fact that one below the rank of professor dared come to him. My list was for current subscriptions, as Doctor [Albert Benjamin] Prescott and I had decided to begin with these and fill up the back numbers later, since we knew that this would take a long time. The reverend librarian growled at the proposed purchase of so many journals in foreign languages and when he looked at the total cost he said with an air of finality and dismissal that it could not be done. I tried to argue and asked him to submit the list to the library committee. This he declined to do and turned me out of his room rudely. This rudeness probably saved my cause, because I am sure that the good old man thought it over and concluded that he had not treated me quite justly. The list which I had submitted carried the annual subscriptions, but many of the journals provided for semi-annual, and some for quarterly payments.

A few days later I faced the good old gentleman again with exactly the same list but with the prices cut down to the smallest time limit, most of the subscriptions being for only three months. The dear old man with no word of apology but with a face as full of kindness as he could mould [sic] it, signed his approval and before my eyes put the list in an envelope and addressed it to the European agency. Then he kindly dismissed me. I could have hugged him but I dared not. I left his room full of elation, tinged and softened with forebodings of what might happen when requests for renewals would come in. Nothing did happen, at least so far as I know, and the journals on that list, so far as war interruptions have permitted, are still coming to the library of the University of Michigan.[28]

The extent of Vaughan's success in this endeavor may be seen in the increase within the library collections. Beginning with a level of 1,500 volumes during the 1870s, the number of holdings increased to 2,626 volumes and 614 pamphlets by 1885. Appropriations were increased to over $1,000 annually, which not only allowed for subscriptions to most medical publications, but, once Vaughan was appointed dean, for the subsequent completion of many of the back issues of the journals.

Among the additional changes to the library system begun during these first years of Vaughan's appointment was a consolidation of the library committee into three members. Previously, appropriations had been divided on a departmental basis; expensive material could be managed only when various departments were willing to carry out a joint purchase. Instead, the three member committee would make the decision for such purchases. Further, expensive back issues would be purchased only on the basis of need; lesser, or rarely used issues could be obtained from the Library of the Surgeon General's Office (later the Army Medical Library and now the National Library of Medicine).[29] George Dock was particularly useful in obtaining historical material for the growing library. During trips to Europe he frequented rare book shops looking for material to be purchased and brought to Ann Arbor.

The challenge facing the medical student in maintaining current knowledge was clearly described by Dock when discussing the history of the medical collection. "The difficulty of getting as many journals as is desirable can be understood when we realize the great number published, and the rate at which they multiply. In 1887 there were 700 published in all parts of the world; now there are about 1,300. Perhaps the rate will not keep up. Weak journals drop out or consolidate with stronger ones with pleasingly increasing rapidity. But it is certain there will be a steady increase of periodicals that no progressive library can afford to ignore. An example of this kind is given by the history of the X-rays. This discovery quickly called forth not only many books but eight or ten periodicals, some of them very costly, and almost all necessary to the practicing physician."[30]

The challenge, both in cost as well as quality, was not unique for the times, as any contemporary librarian would acknowledge. An estimate for the year 2000 placed the number of journals as upwards of one million.[31] Not all these journals are equally significant, of course. Many are available online. But the phrase "more and more about less and less" would certainly apply to the growth of scientific knowledge. In quoting his mentor Osler, Dock emphasized the importance of reading in expanding one's knowledge. "It has been said that 'there should be in connection with every library a corps of instructors in the art of reading, who would, as a labor of love, teach the young idea how to read.' This art should be taught before the young idea begins to study medicine."[32] This would apply as equally in current times as in 1907, when students often forsake reading for internet searches, or reading the plethora of "Wiki" links.

Years later Vaughan was instrumental in founding the *Journal of Laboratory and Clinical* Medicine (1921), serving as the first editor-in-chief of the journal which continues today as one of the most respected of such publications; his son, Dr. Warren Vaughan, later succeeded his father as editor. Vaughan's personal library was eventually donated to the university, adding to collections donated by other prominent faculty over the years. The university medical library still ranks among the most complete in the country well into the twenty-first century.

Chapter 5

Ptomaines and Leucomaines

Vaughan had characterized tyrotoxicon as a product of bacterial putrefaction, a member of a complex of bacterial degradation products classified as either ptomaines or leucomaines.[1] The ptomaines, a term coined some years earlier referring to amide containing compounds produced from putrefying matter, either animal or vegetable, while leucomaines referred to natural substances already found in living tissues and considered normal products of metabolism. During these early years of research into the mechanisms of disease, Vaughan and Novy divided diseases into two classes: "the infectious and the autogenous, those introduced from without [infectious] and those originating within the organism [autogenous]."[2] The discovery of tyrotoxicon fell into the category of ptomaines, and was one of the factors which led Vaughan to suggest disease, or at least infectious disease, was the result of bacterial growth in which production of a chemical poison takes place; according to the theory, the chemical poison in turn is absorbed by the tissues. The characteristic pathology of the disease then follows as a result. Vaughan thus considered that infectious disease originates with an external organism, a germ, which through the putrefaction of chemicals within the body produces the disease. Each germ was believed to produce its

Frederick G. Novy (Bentley Historical Library, University of Michigan [F.G. Novy file]).

own characteristic poison. It followed that treatment of the disease must involve either direct elimination of the infectious agent, or at least the neutralization of the chemical toxin. Failing either of those methods of resolution, the goal would be to keep the human host alive long enough for the disease to resolve itself.[3]

In contrast, Vaughan attributed autogenous disease to "disturbances between tissue metabolism and excretion. They are prevented by keeping these functions of the body in harmony. They are treated by hastening elimination or by retarding or modifying metabolism or by both."[4] Vaughan's theories of disease, particularly from the perspective of a physician trained more in the area of chemistry rather than strictly in the nascent field of bacteriology, provide insight into his thinking and analysis of the subject during this period.

In the first edition of the book *Ptomaines and Leucomaines* (1888) in which Vaughan and Novy laid out their hypothesis on the cause of infectious disease, the authors drew on the recently established rules developed by Koch in associating a specific organism and its role as an etiological agent of a disease; the four rules subsequently became known as Koch's Postulates. "(1) The special bacterium must be present in all cases of the disease." This of course is not by itself indicative of the etiological agent since the presence of the germ may be a consequence of the disease, not the cause. "(2) The special microorganism must be freed from other organisms and from all matter found with it in the diseased animal." In other words the germ must be grown in pure culture in the laboratory. This concept of a pure culture was also a development provided by Koch and his associates during the 1880s. "(3) The special germ, thus freed from all foreign matter, must, when properly introduced, produce the disease in healthy animals." The statement is self-evident. The isolate must, by itself, reproduce the disease. "(4) The microorganism must be found properly distributed in the animal in which the disease has been introduced."[5]

Koch's Postulates, in their theory equating infectious disease to a specific etiological agent, remain to this day a defining feature of medicine. However, even Koch found portions to be problematic. Neither the tuberculosis nor the cholera bacilli, each isolated (or re-isolated in the case of cholera) and identified by Koch, readily grew in non-human laboratory animals. And while at least one human disease had been associated by then with what are now recognized as viruses—Pasteur and rabies—neither the concept of viruses nor the mechanism to grow them in the laboratory would be known for at least another generation.

So assuming, correctly, that germs cause disease, how did Vaughan hypothesize in this early work as to the mechanism? First he addressed other prevailing theories, and playing the devil's advocate, explained why these alternate hypotheses were inadequate, at least as applied to anthrax. The anthrax bacillus had been isolated by the Frenchman Casimir Davaine several decades earlier, but it was Robert Koch who in 1876 firmly established its role as the etiological agent. It had been suggested by Otto Bollinger, a professor at the Munich Veterinary School, that since the anthrax bacillus is aerobic and symptoms of anthrax include those associated with deprivation of oxygen in the animal, it functions by depriving red blood cells of oxygen. Both Davaine and German pathologist Rudolf Virchow had supported this argument by finding large numbers of anthrax bacilli in the blood of infected animals. However, Vaughan recognized there were too many exceptions to these findings to explain the mechanism of anthrax as one associated with oxygen deprivation. Furthermore, even when large quantities of bacilli were present, blood still retained the red color of oxidation; nor did exposure to air containing increased oxygen ameliorate the symptoms. Vaughan concluded that "the theory that germs destroy life by depriving the blood of its oxygen has been found not to be true for anthrax, and if not true for anthrax, certainly it cannot be for any other known disease."[6]

Bollinger also observed that sections of kidney obtained from guinea pigs dead from anthrax contained large quantities of bacilli which formed emboli in the capillaries, causing distention and interfering with the normal function of the organ; he observed similar phenomena in other organs, calling his observations the "mechanical interference theory." Vaughan dismissed this idea by pointing out others had not observed a similar phenomenon.[7]

A third hypothesis as to the mechanism of disease was that bacteria utilize the "proteids" (proteins) necessary for cell function, depriving the cell and causing death. But Vaughan dismissed this idea by pointing out in many cases, death occurs so rapidly that the germ would have insufficient time to consume significant quantities of proteids. Many times the organism is not even in the proper site to contact large quantities of material.[8]

Almost by default, Vaughan was left to conclude that some illnesses are the result of chemical poisons produced by bacteria. In part, this provided the basis for his hypothesis of ptomaine and leucomaine products as the immediate causes of many forms of illness. Once again the specifics of his "Physiological Alkaloid" hypothesis were arrived at by default. Vaughan allowed for three mechanisms by which chemical poisons may underlie disease. First, "the microorganisms themselves may be poisonous, or the poison may be an integral part of them.... In order for the conditions for this theory to be fulfilled, the microorganism must be present in the blood before any of the symptoms appear." But the problem with this hypothesis, as pointed out by Vaughan, was that in many diseases, including anthrax, the organism was not detected in the blood until the host was nearly dead. In many instances no organism could be detected in the blood at all. The German physician Albert Hoffa even tested this directly, injecting large quantities of anthrax bacilli directly into the blood. While the animals did subsequently succumb, symptoms of anthrax did not develop for at least another twenty-four hours and sometimes as long as seventy-two hours after exposure, not the expectation if the organism was the direct source of the chemical poison.[9] It may be noted, however, that Hoffa did subsequently isolate a chemical poison from pure cultures of anthrax as well as from the bodies of animals dying from the disease, a ptomaine he referred to as anthracin. The toxin produced symptoms of the disease when injected under the skin.[10] Hoffa was actually correct in his assumption in the role played by a chemical poison, as we are now aware the symptoms of anthrax are indeed associated with toxin production.

The second possibility considered by Vaughan for the origin of an alleged chemical poison was that "the microorganisms may be intimately associated with or may produce a soluble, chemical ferment, which, by its action on the body, produces the symptoms of the disease and death. This theory formerly had a number of ardent supporters, among whom might be mentioned the eminent scientist [Anton] de Bary. But Pasteur proved the theory false when he filtered anthrax blood through earthen cylinders, inoculated animals with the filtrate, and failed to produce any effect."[11]

Others later attempted similar methods to test for the presence of the toxin, again with negative results. The theory underlying the attempts carried out by de Bary and others to identify a chemical toxin as the cause of the symptoms of anthrax did not significantly differ from that of Hoffa referred to above. Why they failed to observe the presence of a ptomaine, using the contemporary vernacular, is unknown. As we now understand, culture conditions are critical in inducing an organism to express those genes associated with toxin production—that of diphtheria is a prime example—and one may surmise that such requirements had not been met. The fact that Vaughan did not refer to Hoffa's more significant experiments demonstrating production of an anthrax toxin was likely the result either of timing: the first edition of this book was published in 1888, prior to Hoffa's demonstration; or Vaughan may have simply been unaware of this aspect of Hoffa's work.

The third possibility considered by Vaughan for production of the chemical poison was that the bacillus may be "splitting up pre-existing complex compounds found in the body. This theory is supported by analogy, when we remember that the ordinary putrefactive germs produce such chemical poisons, as has been demonstrated by the work of [Danish physiologist Peter] Panum and others. These poisons are ptomaines, and the truth of this theory may now be said to amount to a positive demonstration."[12]

Vaughan continued by explaining his theory of how infectious disease would arise in the organism. "An infectious disease arises when a specific, pathogenic microorganism, having gained admittance to the body, and having found the conditions favorable, grows and multiplies, and in doing so elaborates a chemical poison which induces its characteristic effects."[13] Vaughan provided a number of examples of such ptomaines, both localized as well as systemic, and in doing so attempted to explain the pathology of several important diseases. "In the systemic infectious diseases, such as anthrax, typhoid fever and cholera, this poison is undoubtedly taken into the general circulation, and affects the central nervous system."[14] Vaughan has applied the concept of a bacterial chemical poison as the source of the severe fluid loss associated with cholera. He was correct. But while Koch had suggested the possibility of such a toxin associated with the etiological agent he had isolated in 1884, the reality still remained controversial at the time this was written, a fact Vaughan later acknowledged.

Vaughan certainly would find contemporary support for his interpretations of the role played by bacterial putrefaction in human disease. The experiments of Peter Panum are described above. About the same time Vaughan and Novy were carrying out their experiments on production of toxic derivatives from proteids, the German scientist Ludwig Brieger had isolated alkaloid products of bacterial putrefaction, and after injection into animals, observed the presence of tremors, spasms or diarrhea and other characteristics of chemical poisoning. In some instances the activities of the heart or lungs was disrupted. He determined that the chemical basis for these isolates was likely that of an amide derivative[15]; Brieger designated the poisonous ptomaine as a toxine. Others, including the Italian chemist Francesco Selmi, who isolated ptomaines from human cadavers, found similar results to those reported by Brieger; it was likely Selmi who coined the phrase "ptomaine poisoning."[16]

Vaughan applied his theory to localized diseases as well. "In the local infectious diseases, such as gonorrhea, and infectious ophthalmia, the principal action of the poison has both a local and a systemic effect. Thus, it is by no means certain that the ulceration of typhoid fever is due directly to the bacillus. On the other hand, it is altogether probable that the anatomical changes in the intestine result from the irritating effects of the ptomaine at the place of its formation."[17]

Vaughan used the example of cholera as a disease associated with production of a chemical poison. The theory rested on a series of experiments, albeit indirectly and incorrectly interpreted given the information we now have. "Although the ptomaine of cholera has not been isolated, there are reasons for believing that the comma bacillus of Koch is one of the most active chemically of all known pathogenic organisms. In the first place [Ludwig] Bitter has shown that this germ produces in meat-peptone cultures a peptonizing ferment, which remains active after the organism has been destroyed. It was shown that this ferment, like similar chemical ferments, would convert an indefinite amount of gelatin or coagulated albumin into peptone."[18] Vaughan was correct for the wrong reason. The pathogenic properties of the comma bacillus (now *Vibrio*) are associated with a toxin, albeit one produced locally in the intestine and which acts directly on intestinal cells; Koch as well had found evidence for production of a toxin by the cholera bacillus. The peptonizing enzyme described by Bitter and others was likely the gelatinase which is also produced by the organism.

During these years of the late 1880s experimental evidence lent more support to the theory that the pathological effects found in other diseases were similarly due to production of a bacterial poison. Brieger isolated what he referred to as ptomaines from cultures of the tetanus bacillus as well as from tetanized muscle; inoculation of these compounds into animals produced convulsions; though Koch's Postulates referred to the etiological agent itself, Brieger's experiments certainly conformed to the same concept associating an agent with disease. Alexandre Yersin and Emile Roux in France, as well as Friedrich Löeffler in Germany, all demonstrated that a filtered bacterial product from the diphtheria bacillus likewise produced symptoms of that disease. It would not require a significant shift in thinking to equate such bacterial products, often referred to at the time as ptomaines, to be equated with exotoxins.

Summer Diarrhea in Children

"The contents of the intestines in the so-called summer diarrhea of infancy swarm with bacteria of many species and some of these produce most powerful poisons." Vaughan wrote.[19] It was not unusual for children, particularly infants, to suffer from what was commonly known as summer diarrhea. Broadly defined, it was diarrhea which usually resulted from ingestion of milk or milk products which had been contaminated with bacteria and had been allowed to "incubate" in the warm weather. Since these illnesses were so common during the summer months, in the period before bacteria were determined to be the source of the problem they were sometimes referred to as thermic fevers. A common feature in many of the infants who developed the illness was that they likely would have been bottle fed. In fact Vaughan noted this possibility while emphasizing the importance of nursing the infant. "The child taking its nourishment directly from the breast of the healthy mother obtains its milk practically germ-free, while the one taking cow's milk receives along with this food many kinds of bacteria, some of which are very harmful. These diseases are confined to the summer months because the germs which elaborate poisons in milk require a relatively hot temperature for their growth. During the hot months of summer these bacteria are widely distributed, and easily find their way into milk. They grow rapidly and produce chemical poisons. Furthermore, decomposing matter harbors and supports these bacteria at a time when the outdoor temperature is high enough to allow their growth."[20]

Some of the bacteria Vaughan isolated from these cases resembled, either physically (i.e., bacilli) or biochemically, the cholera bacillus. The diarrhea exhibited by the children appeared as a milder version, if only by comparison, of that found in cholera patients. Consequently physicians sometimes identified these organisms within the categories of "cholera infantum" or "choleriforms." An alternative explanation, given the lesser likelihood of mortality and the observation that infants who are nursed rarely develop the illness, the contaminations were that of the colon bacillus, now known as *Escherichia coli*. Vaughan isolated the by-products of metabolism produced by these bacteria and found these "proteids" to exhibit the toxigenic properties which he had previously identified as being characteristic of other pathogenic bacteria. When he injected the toxigenic material under the skin of test animals, he observed that the animals exhibited symptoms of poisoning, including vomiting, fever and purging (both vomiting and diarrhea).[21]

The difference between cholera infantum, sometimes called cholera morbus, and the summer diarrhea was significant. True cholera was characterized by a sudden onset and rapidly progressed in an infant, usually resulting in death. The source could be either contaminated food or water. Summer diarrhea was generally confined to infants not being breast fed. The tyrotoxicon isolated by Vaughan (see Chapter 3), also a by-product of bacterial putrefaction, was linked by some, apparently including Vaughan himself to cholera infantum.[22] While the term cholera infantum might apply specifically to cholera, its meaning among the contemporaries of Vaughan was more "generic," referring to any of several potentially severe gastric illnesses of infants.

Regardless of the chemical makeup of the toxin, Vaughan was of the opinion that once the diarrhea began in the infant the prohibition of milk must be "absolute. Sterilized milk is not to be thought of and even the breast of the mother or the wet nurse must be denied."[23] The assumption by Vaughan was that the toxin could be a putrification by-product of cleavage of the "proteid," in this case the protein in the milk.

Vaughan himself was certainly aware of the difference between the two agents of diarrhea. "The microorganisms which produce the catarrhal or mucous diarrheas of infancy in summer [i.e., summer diarrhea] may be, and probably are, only putrefactive in character, but those which produce the choleriform or serous diarrhea, true cholera infantum, are more than putrefactive; they are pathogenic; they produce a definite chemical poison, the absorption of which is followed by the symptoms of the disease.

> While I believe that all the summer diarrheas of infants are bacterial in origin, I do not believe that the same bacterium is present in all. The more I observe these diseases clinically, the more I study the subject experimentally in the laboratory, the more thoroughly do I disbelieve in their unity. In the choleriform diarrhea the symptoms come on suddenly and violently. It is true that this affection may occur as an intercurrent disease in the course of an ordinary catarrhal diarrhea, it may be true that it is more likely to occur under these conditions than in a child previously healthy; but when it does appear in such a case the symptoms change.... This disease [cholera infantum] differs from the catarrhal diarrheas of infancy in symptomatology and in pathology, and there are reasons for the belief that there is also a difference in etiology, and if we accept the germ theory for these diseases we must believe that the germs causing them are different, or that the same microorganisms will produce in one child a well-marked train of symptoms and characteristic anatomical alterations, and in another wholly different effects and lesions. The poison of choleriform diarrhea acts immediately on the nervous system, most probably having its chief effect upon the sympathetic nervous system; the most pronounced early effect of the poison of catarrhal diarrhea is a local irritant.[24]

Vaughan was proven particularly prescient. He pointed out that in order to understand the changes which take place in the intestine as a result of infection, it is helpful to understand the contents of the normal flora.

> It now devolves upon the bacteriologist to isolate the microorganisms present in the various forms of summer diarrhea, and upon the chemist to study the putrefactive products of these germs as they are grown in various media. There has already been considerable activity in this line of work among bacteriologists, but until the appearance of the monograph of Escherich, a little more than a year ago, we knew nothing positively or accurately concerning the normal bacterial vegetation in the intestines and stools of infants, nor of the effects of alterations in the food of the child upon these growths. Escherich has done his work so thoroughly that it will serve as a basis from which the pathological organisms may be studied, and we hope for some good results in the near future.[25]

The bacteria which produce these [diarrheal] diseases prove harmful by splitting up complex molecules and forming chemical poisons [ptomaines]. The answer to this question, How do germs cause disease?, has now, thanks to the labors of Selmi, Brieger, Hoffa, and others, has now been satisfactorily answered. They do so by elaborating chemical poisons, some of which are powerful in action, are rapidly absorbed, and manifest their effects upon the central nervous system, as in Asiatic cholera, and, as I believe, in the choleriform diarrhea of infancy; others are less powerful and act primarily as local irritants, inflaming the tissues and leading to necrotic changes, as is most likely the case in the catarrhal diarrheas of infancy.... The ptomaines of tetanus, anthrax, and typhoid fever have been isolated, and I think that in tyrotoxicon we have the chemical poison or ptomaine of genuine cholera infantum. This substance is rapidly formed in milk when the germ is present and the conditions are favorable for its growth. Newton and Wallace, in investigating the outbreak of milk-poisoning at Long Branch in August, 1886, found that the milk drawn from the cows at noon became sufficiently poisonous within six hours to affect seriously those who partook of it; they also found that the toxic agent in milk was tyrotoxicon.[26] The microorganism which produces this ptomaine has not yet been isolated, but we know that the production of this poison is due to the action of a ferment, because a small quantity of poisonous cheese, cream or milk added to good milk will soon render the whole poisonous, if it be kept at the proper temperature.

The following are the reasons for believing that tyrotoxicon is the exciting cause, in at least some instances, of choleriform diarrhea in infants. (a) This ptomaine results from the putrefaction or bacterial fermentation of milk. The disease is confined (with very few exceptions, which are easily explained) to children fed upon cow's milk; (b) Tyrotoxicon has been found in the milk given to a child immediately before the appearance of the symptoms of choleriform diarrhea; (c) The symptoms of the disease increase if the administration of the milk is continued, and abate when milk is withdrawn from the food.[27]

The latter observation is the reason Vaughan recommended that the infant should receive no additional milk, from any source, during the course of the illness. Vaughan's theory that tyrotoxicon, a product of bacterial putrefaction, was the poison which produced the symptoms of cholera infantum was perfectly logical, as noted in his explanation. He was also careful in differentiating the illnesses of choleriform diarrheas and catarrhal diarrheas. While each was characterized by a specific etiological agent as well as a specific ptomaine, he was convinced they represented two different illnesses caused by two different organisms.

Autogenous Diseases

Vaughan in 1888 briefly addressed the presence of autogenous diseases, those which originate within the body, in the final pages of *Ptomaines and Leucomaines*.

5. Ptomaines and Leucomaines　　　　　　　　　　65

It is true, without exception, so far as we know, that the excretions of all living things, plants and animals, are poisonous to the organisms which excrete them. A man may drink only chemically pure water, eat only that food which is free from all adulteration, and breathe nothing but the purest air, free from all organic matter, both living and dead, and yet that man's excretions would contain poisons. Whence do these poisons originate? They are formed within the body. They originate in the metabolic changes by which the complex organic molecule is split up into simpler compounds. We may suppose—indeed we have good reason for believing—that the proteid molecule has certain lines for cleavage along which it breaks when certain forces are applied, and that the resulting fragments have also lines along which they break under certain influences, and so-on until the end products, urea, ammonia, water and carbonic acid gas, are reached: also that some of these intermediate fragments are highly poisonous compounds has been abundantly demonstrated.... While it is true that germs are constantly present in the human intestines, and that they undoubtedly act upon our food, there is no more reason for believing that they are concerned in the actual production of all the leucomaines, than there is for believing that bacteria produce morphine in the poppy.[28]

Vaughan even placed what we now recognize as food allergies or other digestive "abnormalities" into the category of autogenous diseases. "In some persons the tendency to the formation of poisons out of certain foods is very marked. Thus, there are some to whom the smallest bit of egg is highly poisonous [perhaps severe egg allergies resulting in anaphylactic shock]; with others, milk will not agree [lactose intolerance]; and instances of this kind are sufficiently numerous to give rise to the adage, 'What is one man's meat is another man's poison.'"[29]

Chapter 6

War and Disease

The origins of the Spanish-American War are well known and beyond the purposes of our story. Ostensibly, the issue was the independence of Cuba, one of the few remaining possessions of Spain in what had once been a vast empire in the West. Cuban rebels had revolted against Spanish rule several times in previous decades, but it was a combination of American political and economic issues which were leading the United States to likelihood of confrontation with Spain. The event which led to the declaration of war was the destruction of the battleship USS *Maine*, sent to Cuba by President William McKinley to protect American citizens and interests. The explosion in the early morning hours of February 15, 1898, sank the ship, with the loss of 266 sailors. While the cause of the explosion was likely an accident with the boilers, it was immediately blamed on a Spanish mine. That April, Congress declared that a state of war existed between the United States and Spain.

Between the beginning of the war and the official end of hostilities after the capture of Manila in the Philippines by American forces in August 1898—the official peace treaty was signed that December—approximately 273,000 men served in the army. The actual number of fatal battle casualties was relatively low, approximately 369 men. However, over 1,900 men died from disease, largely typhoid fever, but also malaria and yellow fever, both of which were mosquito-borne.

Once war was declared and the United States government called for volunteers, the students in Ann Arbor were among those who enthusiastically responded with patriotic fervor. In response to calls for enlistment, a mass meeting was held in June at University Hall, then located near the site of present-day Mason Hall. Student speakers were surprisingly moderate, urging their peers to wait before joining the war. However, in response to one speaker urging that the ranks of unemployed be the first called to the army, Vaughan himself spoke. "God pity the country whose tramps must fight its battles; it

Siboney, Cuba, as it appeared during Spanish-American War (Images for History of Medicine/National Library of Medicine).

is true that you are here to acquire an education with the purpose of fitting yourself for the work of life; but I would rather see these walls crumble into dust than to see you hesitate to go when your country calls. You have duties towards your parents, but your first duty is to serve your country."[1]

Vaughan himself had considered enlisting. He went so far as to submit a letter to George Sternberg, the surgeon-general, volunteering his services:

> University of Michigan
> Department of Medicine and Surgery
> Ann Arbor, March 26, 1898
>
> Geo. M. Sternberg, M.D., LL.D.,
>   Surgeon-General, U.S.A.
>
> Dear Dr.:
>
>   If there be any need of volunteers for the Medical Service and I am not too old, I wish to be counted in.
>
>   I suppose that you have your hands full just now.
>
>   Yours
>   V.C. Vaughan."[2]

Reed Board investigating yellow fever in Cuba; clockwise from top left: Walter Reed, Aristides Agramonte, James Carroll and Jesse Lazear (Images for History of Medicine/National Library of Medicine).

But with a wife and five children he did not consider it practical. However, his speech had caught the ear of Michigan Governor Hazen Pingree, who the following day contacted Vaughan to inform him that he had just signed a commission placing Vaughan in the army, assigning him as a surgeon, (he later achieved the rank of major), to the 33rd and 34th Michigan Volunteer Infantry brigades. Vaughan acknowledged that while there are many reasons one might enlist, in his case it was because he "talked too much."[3]

After embarking at Alexandria, Virginia, followed by a short stay at Old Comfort Point, Vaughan boarded the cruiser *Yale* and sailed to Cuba. The ship dropped anchor near the village of Siboney on the Caribbean Sea, east of Santiago. In the absence of a harbor, the men of the *Yale*, including Vaughan, disembarked into lifeboats to head for shore. It was here that Vaughan had his first adventure in the "field." About one hundred yards from shore the boat was swamped, and Vaughan and the others were forced to swim to land. Vaughan would spend a significant period of his time while on the island in the vicinity of Siboney.

Early on July 3, the decisive Battle of Santiago began, representing the first and only time Vaughan came under fire. Vaughan's role as a physician entailed the dressing of battle wounds, which included the rescue of a soldier with a wound in the foot, accomplished while under fire. Twenty-six years later (as Vaughan wrote, "Uncle Sam may be slow in conferring honors but he seldom wholly forgets") he received a citation and a Silver Star.[4]

Soon after the battle, Vaughan encountered his first case of yellow fever among the troops. While some few suspected the disease might be transmitted by mosquitoes, the reality was that in 1898 neither the etiological agent—a virus—nor the method of transmission—the mosquito as vector—was established. Since physicians (incorrectly) suspected the disease might be contagious, the solution was to establish isolated yellow fever hospitals; Vaughan and others carried tents and supplies to a site in the mountains where additional victims might be housed. Within a week, several hundred troops had come down with what was often a fatal disease.

Shortly afterward, while treating victims of the mosquito-borne disease as best he could given the restrictions inherent in a war zone, Vaughan himself fell victim to yellow fever, posting a temperature well over 100° F, and repeatedly vomiting the characteristic black substance typical of severe forms of the disease. Some years later, Vaughan learned that doctors had not expected him to survive. After several weeks, during which Vaughan lost over fifty pounds, a quarter of his weight, he was evacuated from Cuba along with some 300 convalescent troops and placed on the ship *City of Santiago* heading to Florida.[5] After spending a short time in Tampa Bay,

Vaughan was ordered to proceed to New York, arriving in that city in mid-August.

The Typhoid Commission

Between May and September 1898 nearly 21,000 cases of typhoid fever were reported among the troops stationed within the United States; approximately 1,600 of these men died. The actual incidence of typhoid was likely higher, as these represented only those which were reported; additional outbreaks of illness were almost certainly misdiagnosed, as with the cases of typhomalaria. Typhoid accounted for nearly 90 percent of the total deaths from disease.[6]

The cause of the outbreaks was perfectly clear, or even in retrospect should have been, to those aware of the need for fresh, clean water. The experiences of the Seventh Corps, originally stationed in Tampa, Florida, prior to embarkation to Cuba, were typical. The corps consisted of three divisions, a total of 31,000 men. That May the corps was divided into two parts, with one division sent to Miami and the other two divisions sent to Jacksonville. A week later the first cases of typhoid appeared in Jacksonville the first week of June, likely brought by soldiers recently arriving from Virginia. By the end of July dozens more cases had been diagnosed; by December over 400 likely cases of typhoid had developed within a 1,200 man regiment. Other regiments suffered in a similar manner, with between 20 and 30 percent of the men being affected.[7]

The probable cause of the outbreaks could be traced to the means by which fecal waste was being disposed. Like many encampments, disposal utilized the tub system to clean the latrines. Waste was collected in old whiskey barrels the contents of which were then dumped into larger barrels being carried on wagons. Both urine and fecal waste were collected in the mixture, and barrels were often filled to overflowing. One historian explained: "The disposal of fecal matter from the sinks was by hauling in tubs, filled to overflowing and constantly slopping and spilling along this (shell) road within 20 feet of our tents, and generally at mealtime. It was impossible to keep the dust out of our food, and the passing of these filthy and vile-smelling wagons while the men were eating was almost in itself cause enough for sickness, even had infectious germs been absent from the loathsome loads."[8]

Ultimately, typhoid fever became the primary killer of soldiers in the war; morbidity of the disease was even higher than it had been during the beginning of the Civil War a generation earlier. The outbreak of typhoid in the camps was particularly surprising given the state of knowledge of disease in general and typhoid in particular by 1898. The etiological agent of the disease had been isolated and characterized over a decade earlier, and little controversy existed as to the microorganism's specific role. Furthermore, the contagious nature of the disease and its transmission by means of fecal contaminated food or water had been well established.

By 1896 French physician Fernand Widal had developed a diagnostic test for the disease, based upon the ability of blood serum isolated from typhoid patients to agglutinate the typhoid bacillus; the procedure became known as the Widal test. Even though the typhoid bacillus bore some resemblance to the more common colon bacillus (*Bacterium coli*), in theory a trained bacteriologist at least should have been capable of isolating and identifying the organism.

One problem, however, was that few medical officers had been trained in such bacteriological techniques. Diagnosis of typhoid was based primarily on how the disease was presented—for example, high fever—and its response to the malarial treatment of administration of quinine. If the symptoms did not abate after treatment, the disease could not be malaria.[9]

Not surprisingly, the high incidence of typhoid among the soldiers created an uproar within the general public as well as among the troops themselves. As a consequence, Surgeon General George Sternberg on August 18 announced the appointment of a three man board, known as the Typhoid Board, to investigate. Sternberg had been appointed to his position in 1893. By this time he was considered the preeminent bacteriologist in the country. His practical medical career had begun when serving as an assistant surgeon in the Civil War, during which he had been cited for his service, which also included surviving a bout of typhoid. His text *A Manual of Bacteriology* (1892) was considered the definitive

Victor Vaughan, 1898 (Bentley Historical Library, University of Michigan [Victor Vaughan file]).

work on the subject. The Typhoid Board appointed by Sternberg consisted of Major Walter Reed, serving as chairman, Brigade Surgeon Edward Shakespeare and Major Victor Vaughan:

War Department
Adjutant-General's Office,

Special Orders No. 194   Washington, August 18, 1898

A board of medical officers, to consist of Maj. Walter Reed, surgeon, U.S. Army; Maj. Victor C. Vaughan, division surgeon, U.S. Volunteers, and Maj. Edward O. Shakespeare, brigade surgeon, U.S. Volunteers, is appointed to meet in this city at the earliest date practicable for the purpose of making an investigation into the cause of the extensive prevalence of typhoid fever in the various military camps within the limits of the United States, under such instructions as it may receive from the Surgeon-General of the Army. The board will call the attention of the proper commanding officers to any insanitary conditions which may exist at the camps visited by it, and will make recommendations with a view to their proper correction. The report of the board will be forwarded to the Surgeon-General as soon as practicable after the completion of the investigation contemplated.

Such journeys as may be required under the above order are necessary for the public service.

By order of the Secretary of War:

H.C. Corbin, Adjutant-General[10]

Walter Reed would someday be better known for his work in confirming the role of the mosquito in transmission of yellow fever. He was born in Virginia in 1851, the son of a Methodist minister. He received his degree in medicine from the University of Virginia, having spent two years of study at that institution. Having graduated at age 18 in 1869, Reed is still the youngest person to receive a degree from that medical school. Following graduation Reed entered Bellevue Hospital Medical College in New York, from which he earned a second medical degree in 1870. After interning in several hospitals in that city, Reed was appointed in 1873 as assistant sanitary officer for the board of health

George Sternberg (Images for History of Medicine/National Library of Medicine).

in Brooklyn, then a city separate from that of New York. Two years later he enlisted in the army, and after passing a qualifying examination, was commissioned as a first lieutenant at the local military base. During the next several years, Reed was transferred between a variety of posts, many of them in the far West.

In 1889 Reed was posted in Baltimore, and taking advantage of the proximity to the Johns Hopkins School of Medicine, he continued his coursework in the areas of bacteriology and pathology. Four years later Reed was appointed as a professor of bacteriology at the George Washington School of Medicine and as a professor of bacteriology and clinical microscopy at the Army Medical School in nearby Washington, D.C.

Consequently, by the time of Reed's appointment to the Typhoid Commission in 1898, he already had extensive experience in the area of public health, both as a result of having served on the New York Board of Health soon after receiving his degree and having been posted as medical officer and surgeon at army forts throughout the West the previous decade.

Edward Shakespeare's original training had been in the field of ophthalmology, but he subsequently branched out into research and scientific investigation; he was among the first in that area of medicine to apply cocaine as an anesthetic for the eyes. Shakespeare had been born in 1846 in Delaware, graduating from Dickinson College in Carlisle, Pennsylvania. He then enrolled in the University of Pennsylvania Medical School and received his degree in 1869. Three years later he became a member of the staff at Philadelphia General Hospital, with subsequent promotions to positions of pathologist and staff bacteriologist, partially the result of his brief interaction with Robert Koch several years earlier. Shakespeare became well known as a result of his investigation into the typhoid outbreak in the town of Plymouth, just outside Philadelphia, during the mid–1880s. Shakespeare determined the outbreak was the result of fecal contamination of the local water supply by a single victim of the disease. In 1885 Shakespeare was appointed to a commission investigating a cholera outbreak in Spain and Italy. His extensive description of that epidemic, *Report on Cholera in Europe and India* (1890), established Shakespeare as an authority on that subject as well.

The first of the camps inspected by the three members of the typhoid board was that of Camp Alger at Dunn Loring, Virginia, on August 20. An outline of the procedure followed by the board members was provided in Vaughan's autobiography.[11] The first challenge was locating a microscope, not one of which was present in the camp. Further, nobody was immediately available who had experience with the Widal test, necessary to confirm a case of typhoid. After appealing to Surgeon General Sternberg, they were told to "draft" any medical personnel they might need, with cost no object. The physicians they chose were among the most experienced in the country: Drs. William Gray and James Carroll from the Army Medical Museum, Dr. George Dock, recruited from the medical school at the University of Michigan, Dr. J.J. Curry and Dr. Charles Craig.[12]

They spent six days at the site, investigating both typhoid and what other physicians had diagnosed as malaria (typhomalaria); diagnosis by the board was based upon either the presence of the malarial parasite or a positive Widal (typhoid agglutinin) test. It was determined that despite the diagnosis of malaria or typhomalaria by physicians in the camp, nearly all cases of illness were actually typhoid fever.[13]

In fact, despite increasing evidence to the contrary, medical officers at a number of the camps continued to diagnose the illness as typhomalaria. In the past such an error might have been excusable. Some symptoms were common to both forms of illnesses: high fever, muscle or abdominal aches and diarrhea. But by the latter half of the 1890s, diagnostic tools, including the Widal test but also isolation and identification of the etiological agent, could readily differentiate between the two types of illnesses. Sir William Osler, professor of medicine at Johns Hopkins and among the founders of the Johns Hopkins School of Medicine and one of the most notable practitioners of medicine at the time, had written and lectured extensively for years on the differentiation of diseases such as malaria, typhoid and typhus.

In an attempt to put to rest the alternative (and incorrect) diagnosis, when the typhoid board arrived at the camp in Jacksonville, they spoke with the commanding officer, General Fitzhugh Lee—son of the Confederate commander Robert E. Lee—requesting that medical personnel select fifty cases of what they diagnosed as typhomalaria. The men were sent to Fort Myer outside of Washington, D.C., where they were examined by James Carroll. Carroll confirmed the typhoid fever diagnosis. An additional one hundred and fifty cases were sent to a variety of hospitals; all confirmed the same diagnosis of typhoid. In the face of such evidence diagnosis of typhomalaria ceased.[14]

The investigation of the outbreak in Camp Cuba Libre near Jacksonville was also critical in defining the source of the contamination. Prevailing wisdom was that typhoid is generally spread through fecal contaminated water. However, the board observed that water used both by the troops as well as the citizens of Jacksonville was obtained from a series of four artesian wells, ranging in depth from "six hundred and thirty to one thousand and twenty feet," far too deep for simple contamination from latrines.[15] Furthermore, though the population of both the city and the camp was approximately

30,000, typhoid was nearly entirely confined to outbreaks among the troops. The members of the board observed fecal material was being disposed of through the use of a tub system; this explained the likely source for the typhoid bacillus. The result was that contaminated material was dispersed throughout the camp. When the use of these tubs as a means of transporting material from the latrines was eliminated and replaced with lime disinfectant of material, the outbreaks ceased.

The situation as it developed at Camp George Thomas at Chickamauga Park was similar. The camp held as many as 60,000 men, members of the First and Third Corps, at one time; it was estimated that an average of 9.4 tons of feces and 21,000 gallons of urine were produced each day, most of which was disposed of as described above.[16] The camp had been laid out on what had been a bloody Civil War battlefield some thirty-five years earlier. The presence of Chickamauga Creek provided drainage. Surgeon General George Sternberg had included strict instructions about proper sanitary procedures in his first military circular, issued on April 25. Encampments were to be established on high ground and were to allow for proper drainage. Each company would be responsible for its own latrines. These areas were to be treated three times each day with either quicklime or ash and covered with dirt. Anything remaining from the kitchen was to be burned rather than buried. Drinking water was to be boiled before use. Obviously this would apply to the traditional coffee or tea prepared by soldiers at rest, but also applied to any drinking water. Medical officers as well as the ordinary soldiers received frequent reminders to maintain sanitary conditions.[17]

Unfortunately, as is usually the situation when troops are sitting around waiting for orders, the men became bored and began ignoring the sanitary precautions. Latrines were no longer covered as required by orders. In fact the latrines were often simply shallow pits, sometimes placed above the soldiers' tents. Soldiers often failed to boil their drinking water, some with the belief that orders to do so infringed on their personal rights. When the summer rains arrived in late June the results were as one might expect. Streams overflowed and latrine pits were quickly flooded, spreading fecal material throughout the encampment.

The example of the Eighth Massachusetts illustrates the disregarding of common sense by the ordinary soldier. "A spring adjacent to an outpost near the Chickamauga Creek was flooded during the storms, leaving behind a coating of slime after the creek subsided. Acting upon the suggestion that germs floated, 'they plunged a canteen to the bottom of the spring, with the opening stopped by the thumb against the entrance of bacteria. When the canteen was on the bottom, the thumb was removed until the canteen was filled, the opening was then plugged with the thumb, and the supply brought to the surface.' Every person who drank from the spring came down with typhoid shortly afterward."[18]

In addition to the flooding which wreaked havoc with poorly constructed sanitary systems, the water attracted another nuisance: flies. As described by Wintermute, flies were everywhere in massive quantities, literally covering the food. The heat of a Southern summer ensured that any food not properly stored would quickly become spoiled. Hospitals were often little help in dealing with the sick. In the beginning, each regiment established its own hospital facilities. But hospital personnel often assumed that anyone reporting sick was simply a malingerer, and was treated as such. It was not until hospitals were established at the division level that a solution was provided. Eventually Camp George Thomas had to be abandoned.[19]

Between August 20 and September 30, Vaughan, Reed and Shakespeare visited several volunteer army encampments: Camp Alger, Fernandina, Florida; Camp Cuba Libre at Jacksonville; a camp at Huntsville, Alabama; Camp George Thomas at Chickamauga Park, Georgia; the camp at Knoxville, Tennessee; and Camp Meade at Harrisburg, Pennsylvania as well as a quarantine encampment at Montauk Point and general hospitals at Fort Monroe, Virginia, and Fort McPherson, Georgia.[20] At Jacksonville, the colonel in charge of the 3rd Nebraska regiment was William Jennings Bryan, a survivor of typhoid and the nominee for the presidency by the Democratic Party during the previous election; Bryan would be the nominee twice more but would be remembered by posterity largely for his role in the Scopes "Monkey Trial" in 1925.

Travel between the camps included accommodations as comfortable as the government (and Southern Railroad) would allow. The board had access to a private railway car which included a kitchen, along with a chef and servants. Each member of the board had his own private bedroom, while the front of the car had room for an office manned by a stenographer.[21]

The investigation carried out by the board can be considered a "landmark study" with respect to the details into understanding the basis for the typhoid outbreaks. Studies went beyond the obvious into layout and placement of latrines in conjunction with kitchens and water supplies, the levels of personal hygiene, and food storage or preparation. All hospitalized patients were tested to determine whether they indeed suffered from typhoid, or, were diagnosed with malaria; in nearly all cases the actual illness was typhoid. Illnesses which had been previously misdiagnosed as typhomalaria were in all cases determined to be typhoid.[22]

The initial abstract of the board's report was published in 1900, a year after Vaughan had been mustered out of the service; Vaughan continued to contribute to the report until January 31, 1900, at which time Reed and Shakespeare would produce the final copy. However on June 1, 1900, Major Shakespeare died suddenly from a heart attack at the age of 54. Shakespeare had been troubled by heart disease for some time prior to his death, and the pressure and volume of work he was dealing with certainly contributed to his illness. Reed as well died suddenly two years later on November 23, 1902, from a burst appendix; at the time Reed was involved with investigating the role of the mosquito in transmission of yellow fever. It was left for Vaughan to continue with the still incomplete report of the findings by the commission.

As the only surviving member of the typhoid board, Vaughan by default had to complete the report in a timely fashion. It would require another two years—until 1904. The final work would be published in two volumes: the first volume consisted of the actual report, including the board's findings and conclusions; the second volume consisted of tables and graphs illustrating data obtained from each of the camps and regiments—ninety-two, comprising nearly 108,000 men.

The report included fifty-seven specific determinations. The findings included the fact that "typhoid fever existed in every single regiment in the Regular and Volunteer army assembled in the Eastern United States, including the expeditionary force sent to Cuba. Typhoid appeared within every one of the volunteer regiments within eight weeks of their arrival at one of the assembling camps ... typhoid fever was so widespread throughout the country, it was inevitable some would arrive at the encampments already sick."[23] Vaughan estimated that in an average size regiment of some 1,300 men, it was likely between one and four men were already infected with the typhoid bacillus.[24] By arriving at these conclusions, Vaughan was able to address the hypotheses previously presented by Charles Murchison that rather than being associated with a specific infectious agent, typhoid is the result of a putrefaction process taking place in the soil, what had been called by some as the "pythogenic theory": the idea that "the specific poison of [typhoid] fever is generated by the process of common putrescence."[25] More specifically, "Typhoid fever may be generated independently of a previous case by fermentation of fecal, and perhaps other forms of organic matter."[26]

In Vaughan's words, "Translated into the terms of modern medicine, this theory is founded upon the belief that the colon germ may undergo a ripening process by means of which its virulence is so increased and altered that it may be converted into the typhoid bacillus, or at least may become the active agent in the causation of typhoid fever. Many French, English and American army medical officers believe that typhoid fever may originate in this way. Rodet and Roux, of the French army, have stated their belief that outside the body the colon bacillus acquires 'typhogenic' properties. Surgeon-General Davies, assistant professor of hygiene in the English Army Medical School, has expressed his belief in this theory."[27] Vaughan addressed as well the idea that the colon bacillus may undergo a conversion. "There is, therefore, no necessity of resorting to the theory that the colon bacillus may be converted into the typhoid bacillus. Moreover, all the known facts of experimental bacteriology are at variance with this theory."[28]

Vaughan and the other (late) members of the board explained in a clear manner exactly why such alternative theories could simply not be valid, explanations "self-evident" in their words.

> Typhoid fever is more likely to become epidemic in camps than in civilian life because of the greater difficulty of disposing of the excretions from the human body.... The influence of the introduction of sewers into cities in decreasing sickness from this disease is well known to every student of sanitary science. Moreover, since the disease is disseminated by the transference of the excretions of an infected individual to the alimentary canal of others, it must follow that the more thoroughly and completely the excretions are removed from the vicinity of habitations the less will be the danger of infecting the inhabitants. In fact, the whole question of the prevention of typhoid fever in armies is largely one of the dispositions of the excretions.... The elimination of typhoid bacilli from the bowels probably begins soon after infection. If this be true, during the entire period of incubation an individual may be a source of danger to others. Moreover, in most instances of typhoid fever the disease is not recognized during the prodromal [early largely asymptomatic] phase, and during this time the excretions may be laden with typhoid bacilli. It must be evident from this that the only way in which typhoid epidemics can be with certainty prevented in armies is by the complete disinfection of the excreta of all, both the sick and well.[29]

Description of the fecal contamination at Chickamauga was perhaps the most vivid, if not largely typical of the situation at other camps. As noted above, some 9.4 tons of feces were generated daily there. "In the camp of the Third United States Voluntary Cavalry we found the sinks full to the top with fecal matter; soiled paper was scattered about the sinks, and the woods behind the regimental camp was strewn with fecal matter. The Second Kentucky Volunteer Infantry was located in the woods; fecal matter was deposited around trees and flies swarmed over these deposits not more than 150 feet from the company mess tents; the odor in the woods just out of the regimental lines was vile. In the Ninth New York we found three battalion sinks supposed to have been filled with straw and burned out that morning. Fecal matter was found deposited on the ground around trees and a vile odor permeated the air around the sinks."[30]

The commission also suggested the common fly, *Musca domestica*, might in addition serve indirectly as a vector for transmission of the typhoid bacillus. It was no surprise of course, given the lack of proper sanitation procedures at the camps which the commission inspected, that flies were ubiquitous both in the fecal matter as well as in the kitchens where food was being prepared. A "natural" experiment even demonstrated the potential for flies to carry the bacillus. When lime had been sprinkled over fecal matter, flies were observed walking over the waste material, with their feet whitened by the lime which had adhered to them. "It is possible for the fly to carry the typhoid bacillus in two ways. In the first place, fecal matter containing the typhoid germ may adhere to the fly and be mechanically transported. In the second place, it is possible that the typhoid bacillus may be carried in the digestive organs of the fly and be deposited with its excrement...."[31] Flies swarmed so numerously that the first droppings of fecal matter were often covered with them before the act of defecation was completed."[32]

Confirmation of the possible role of the housefly in transmission of typhoid came about during the interim in which Vaughan was compiling the report. In a study of the outbreak of typhoid fever in Chicago in 1902, Dr. Alice Hamilton, often considered the founder of occupational medicine and a former student in the University of Michigan Medical School, was able to isolate typhoid bacilli from houseflies which had fed upon fecal material contaminated with the organism.[33] By the time Hamilton published her findings the following year, both Reed and Shakespeare were dead.

The final report which was completed and submitted in 1904 included both the assignment of "blame" for the extent of the typhoid outbreak as well as recommendations to avoid an equivalent problem in the future. The immediate cause of the outbreak was improper disposal or treatment of the enormous quantities of excrement generated by the troops. The presence of even one or two men carrying the disease would result in the rapid disbursement of the typhoid bacillus. Blame was placed on inexperienced officers who, in hopes of generating favor among the troops they commanded, were often lax in enforcement of common sanitary procedures.

> Regiments were [often] improperly located from a sanitary standpoint. This was done by superior line officers, and sometimes in the face of protests from the medical officers. We have also seen that requests for change in location were disregarded, and regiments were allowed to occupy one site for too long a time. In general, the camps became very filthy. It must, therefore, be admitted, it appears to us, that line officers were to some extent responsible for the condition of the camps under their command. The medical officer can only recommend; the line officer can command. We think it unfortunate that hygiene is not taught in our national military school. It does seem that line officers should be able to recog-

nize the importance of reasonable requests and recommendations made by the medical officers.

"In our opinion it is of the greatest importance that more authority be granted medical officers in all matters pertaining to the hygiene of camps."[34] Additional recommendations were included in the report. "In permanent camps where water carriages cannot be secured ... galvanized-iron troughs containing milk of lime be used for the reception of all fecal matter and urine, and that the contents of these troughs be removed daily by means of the portable odorless excavator. We are aware of the fact that this method of disposing of fecal matter will be attended by increased cost, but we are confident that it will lessen greatly the number of cases of typhoid fever."[35] Cost was at the heart of this issue. Vaughan later pointed out that upon declaring war, Congress had appropriated fifty million dollars for "national defense." The qualifier here was that the money could only be used in that manner; consequently the money was used primarily for establishment of coastal defenses in the event of a Spanish attack. No money was allocated for the medical department.[36]

In the report, the commission estimated that among the 107,973 men and officers who comprised the 92 regiments, an estimated 20,738 cases of typhoid had broken out, representing some 20 percent of the troops. The actual number, if one included unapparent or mild cases, was likely significantly higher. Approximately half the cases were properly diagnosed by the army medical personnel; most of the rest had been incorrectly diagnosed by army surgeons as either malaria or typhomalaria. Of those diagnosed, 1,580 died from the disease, representing 7 percent of the cases.[37]

Many of the recommendations submitted by the commission were subsequently carried out—certainly by the time the United States entered the First World War two decades later. The first relatively crude, by modern standards, typhoid vaccine was also available by that time. The results of implementing the recommendations were clear: among the approximately four million American troops in that war, only 1,529 admissions to military hospitals were attributed to typhoid.[38]

The Typhoid Background: History

That dealing with the outbreak of typhoid fever was a significant challenge during the war was not a surprise. It had been the bane of both military and civilian populations through thousands of years of recorded history. While outbreaks certainly had taken place previously, the earliest "validated" epidemic is currently believed to have been that described by the Greek historian Thucydides in his account of the Plague of Athens during the Peloponnesian War between Athens and Sparta (ca. 430–426 BCE). As described by Thucydides,

> In the first days of summer the Lacedaemonians [Spartans] and their allies, with two-thirds of their forces as before, invaded Attica, under the command of Archidamus, son of Zeuxidamus, king of Lacedaemon, and sat down and laid waste to the country. Not many days after their arrival in Attica the plague the plague first began to show itself among the Athenians. It was said that it had broken out in many places previously in the neighborhood of Lemnos [island off the coast of Greece] and elsewhere; but a pestilence of such extent and mortality was nowhere remembered. Neither were the physicians at first of any service, ignorant as they were of the proper way to treat it, but they died themselves the most thickly, as they visited the sick most often; nor did any human art succeed any better.... It first began, it is said, in the parts of Ethiopia above Egypt, and thence descended into Egypt and Libya, and into most of the King's country. Suddenly falling upon Athens, it first attacked the population in Paraeus—which was the occasion of their saying that the Peloponnesians had poisoned the reservoirs, there being as yet no wells there—and afterwards appeared in the upper city, when the deaths became much more frequent. All speculation as to its origins and its causes, if causes can be found adequate to produce so great a disturbance, I leave to other writers, whether lay or professional; for myself, I will simply set down its nature, and explain the symptoms by which perhaps it may be recognized by the student, if it should ever break out again. This I can the better do, as I had the disease myself, and watched its operation in the case of others.
>
> That year then is admitted to have been otherwise unprecedentedly free from illness, and such few cases as occurred all determined in this. As a rule, however, there was no ostensible cause; but people in good health were all of a sudden attacked by violent heats in the head, and redness and inflammation in the eyes, the inward parts, such as the throat or tongue, became bloody and emitting an unnatural and fetid breath. These symptoms were followed by sneezing and hoarseness, after which the pain soon reached the chest, and produced a hard cough. When it fixed in the stomach it upset it; and discharges of bile of every kind named by physicians ensued, accompanied by very great distress. In most cases also an ineffective retching followed, producing violent spasms, which in some cases ceased soon after, in others much later. Externally the body was not very hot to the touch, nor pale in its appearance, but reddish, livid, and breaking out into small pustules and ulcers. But internally it burned so that the patient could not bear to have on him clothing or linen even of he very lightest description; or indeed to be otherwise than stark naked. What they would have liked best would have been to throw themselves into cold water; as indeed was done by some of the neglected sick; who plunged into the rain-tanks in their agonies of unquenchable thirst; though it made no difference whether they drank little or much. Besides this, the miserable feeling of not being able to rest or sleep never ceased to torment them. The body meanwhile did not waste away so long as the distemper was at its height, but held out to a marvel against its ravages; so that when they succumbed, as in most cases, on the

seventh or eighth day to the internal inflammation, they had still some strength in them. But if they passed this stage, and the disease descended further into the bowels, inducing a violent ulceration there accompanied by severe diarrhea, this brought on a weakness which was generally fatal. For the disorder first settled in the head, ran its course from thence through the whole of the body, and, even where it did not prove mortal, it still left its mark on the extremities.

By the time the two waves of the epidemic had run their course, approximately one-third of the population of Athens had perished, including one-fourth of the Athenian army and its leader, Pericles.[39] Many of the victims were hurriedly buried in mass graves. There is no record of a corresponding outbreak among the Spartan army which was besieging Athens during these years.

Historians have speculated that the etiological agent of the epidemic was any of a wide variety of diseases, ranging from influenza to smallpox, typhus, typhoid or bubonic plague—even Ebola. Portions of Thucydides' description could be applied to any one of these, or just as possible, to none of them. The mystery may have been solved some 2,400 years later.

During 1994 and 1995 excavations in Greece at the ancient cemetery of Kerameikos, a portion of which lay within the walls of ancient Athens, uncovered a mass burial site containing the skeletal remains of some 150 persons and which could be dated to the time of the Athens plague. Some of the skeletons had retained intact teeth, which allowed for the analysis of microbial genetic material which might have entered the dental pulp. Analysis of the microbial DNA resulted in the identification of *Salmonella*, providing strong support for the theory that the plague was an epidemic of typhoid fever.[40]

Typhoid fever and typhus, though two distinct diseases with different etiological agents, were historically linked well into the nineteenth century. The term typhoid itself refers to "resembling typhus," a designation derived from the Greek *typhos*, or pungent odor.[41] The difference between the two illnesses, typhoid fever and typhus, is credited to John Huxham, who in his classic work *Essay on Fevers*, the first edition of which was published in 1739, identified two distinct diseases as "putrid malignant fever (*febris putrida*)" and "slow nervous fever (*febris nervosa lenta*)."[42] The term typhoid fever (*fievre typhoid*) was likely coined by the French physician Pierre Charles Alexandre Louis during his study of the pathological lesions associated with the disease. His work, published in 1829, did not distinguish typhoid fever from typhus, probably because of the dearth of cases in Paris with the latter disease.[43] Louis was not the only French physician to study the disease during this period, but the relatively few cases of typhus in France likewise proved a challenge in diagnosis to others as well. In contrast to the situation in France, typhus was common in Great Britain, while unlike the situation in France, typhoid was relatively uncommon there. Few comparisons were made between the two diseases in England, and French physicians believed the two diseases were actually identical.

Perhaps the first physician to recognize the difference was William Gerhard. Born in Philadelphia, Gerhard was trained in medicine at the University of Pennsylvania, from which he graduated in 1830. He spent the following year working under the tutelage of Pierre Louis in Paris, where he was able to observe cases of typhoid fever. Prior to returning to the United States in 1833, Gerhard observed an outbreak of typhus while he was in Edinburgh visiting the medical school. That and his later observations of the disease in Philadelphia convinced Gerhard that the typhus in Edinburgh and the outbreak in Philadelphia were the result of the same disease, while the illness was not identical to the typhoid described by Louis. In an 1837 publication, Gerhard noted the differences both in transmission of each disease—typhoid was rarely directly transmitted from person to person, while typhus was contagious—as well as differences in the pathological lesions. Patients with typhoid fever commonly exhibited lesions in the intestinal Peyer's patches and mesenteric nodes, while such changes were not found when carrying out autopsies of persons who died from typhus.[44]

The differences in features between typhoid fever and typhus described by Gerhard in his work were confirmed by several French and English physicians in the years immediately following the publication. The British statistician Lemuel Shattuck from Boston likewise noted the differences in epidemiological features of the diseases in compiling the vital statistics for the city; Shattuck, however, while a local merchant and founder of the American Statistical Association, was not a trained physician. It was the British physician Sir William Jenner who produced what is now a classic work in differentiating the epidemiological characteristics of the two diseases. Using careful observations of 1,000 patients admitted to the London Fever Hospital between 1849 and 1851, sixty-six of whom died from "continued fever," as well as his personal experience having survived infections by each of the diseases, Jenner showed outbreaks of the diseases did not take place in an identical manner, and that survival of the patient from one of the illnesses provided no protection against the other. Jenner's differentiation of the pathological changes which took place was particularly telling. For example, his observations of the small intestine and mesenteric glands as well as the large intestine upon autopsy revealed the following: "The presence or absence of lesions of these organs [small intestine and mesenteric glands] was the ground on which the cases of typhoid and typhus fever here analyzed were divided from each other,—consequently they were invariably diseased in one [typhoid] and normal in the other [typhus]. After death from typhoid fever the mucous membrane of the large intestines was found ulcerated in rather more than a third of twenty cases. In no instance after death from typhus fever." His conclusion was typhus and typhoid were not variations of the same disease.[45]

The fecal-oral method of transmission of typhoid would remain controversial until at least development of the germ theory of disease in the 1870s and 1880s. However, in 1856 British physician and epidemiologist William Budd produced the first of several publications which dealt with the nature of transmission of typhoid fever. Budd had previously investigated an outbreak of the disease in the Welsh town of Cowbridge in 1853. Following a ball, over 140 persons developed typhoid fever. Budd found that previous to the outbreak, a visitor to the site had become ill with the disease, and fecal material had been disposed of near the site of the local well.

Particularly telling was Budd's later description of the Thames River as the source of the contagion. (Budd considered the nature of contagion as referring to transmission of disease between individuals regardless of the mechanism.) During the first decades of the Reign of Queen Victoria, 1830s–1850s, the size of London had grown nearly two-thirds and to a population of some three million persons; most of the sewage they generated flowed into the Thames River. The summer of 1858 proved particularly hot, and the effect on the sewage pervading the river can only be imagined in present day. At the same time, unrest in the British colony of India was also an issue before the British Parliament, which now had to address both a possible revolt and the odors which disrupted their meetings. Newspaper headlines read "India is in Revolt and the Thames Stinks."

> For the first time in the history of man, the sewage of nearly three million people had been brought to seethe and ferment under a burning sun, in one vast open cloaca [sewer] lying in their midst. The result we all know. Stench so foul, we may well believe, had never before ascended to pollute this lower air. Never before, at least, had a stink risen to the height of an historic event. For many weeks the atmosphere of Parliamentary Committee-rooms was only rendered barely tolerable by the suspension before every window, of blinds saturated with chloride of lime, and by the lavish use of this and other disinfectants. More than once, in spite of similar precautions, the law courts were suddenly broken up by an insupportable invasion of the noxious vapors from the river. The river steamers lost their accustomed traffic, and travelers pressed for time often made a circuit of many miles rather than cross one of the city bridges.... At home and abroad the state of the chief river was felt to be a national reproach. "India is in Revolt and the Thames Stinks" were the two great facts coupled together by a distinguished foreign writer, to mark the climax of a national humiliation. But more significant still of the magnitude of

the nuisance was the fact that five million pounds were cheerfully voted to provide the means for its abatement [British Royal Commission on Sewage Disposal].[46]

Budd wrote, "The great fact remains that sewers are the principal channels through which this fever is propagated, the proof from all sides is overwhelming that they are so not because of their being receptacles of decomposing organic matter, but solely due to their being depositories of the specific discharges of persons already infected."[47]

The identification of the etiological agent of typhoid has been briefly addressed in a previous chapter. While Karl Eberth is generally given credit for identification of the typhoid bacillus in 1880, both Robert Koch and Edwin Klebs have also been credited with similar findings; neither convincingly demonstrated at the time that their isolate was the actual etiological agent. Klebs in particular could win the argument of priority. In April 1880, three months prior to the date Eberth presented his results, Klebs reported the presence of both short bacilli as well as larger filamentous forms in the Peyer's patches and mesenteric lymph nodes of twenty-four victims of typhoid, results identical to that soon after reported by Eberth; none were observed in these sites among cases of other diseases. Furthermore, Eberth reported the presence of spores in his description of the bacilli, though this was likely an error due to the staining procedure.[48] However, if simply the identification of the rod-shaped organism in intestinal tissue of typhoid victims is the major criterion for discovery, that honor would more properly be given to the Polish pathologist Tadeusz Browicz. In an article entitled (and translated) "Vegetable Parasites in Typhoid Fever," Browicz described short rod-shaped organisms in the organs of typhoid victims whom he autopsied. The organisms were isolated and grown in culture, demonstrating their viability. Browicz observed similar organisms in the feces of persons hospitalized with typhoid fever. As with Klebs, Browicz never demonstrated the actual pathogenic potential of the organisms which he had observed.[49]

Priority for growing the typhoid bacillus is generally given to Georg Gaffky in 1884. Here again the credit arguably might belong to Klebs, who three years earlier inoculated gelatin cultures with isolates from the mesenteric lymph nodes of a typhoid victim and reported the growth of rodlike organisms. Gaffky's primary objections to Klebs' assertion was that the latter failed to observe filamentous forms present in some tissues. The use of liquid cultures by Klebs also made it difficult to determine whether he had actually obtained a pure culture; Gaffky grew his isolates on a solid medium.[50]

Certainly by the time of the Spanish-American War it was clear how the disease was transmitted—fecal pollution—whether one was fully accepting of the germ theory or not. An accurate test for identification of typhoid fever in the patient had also been developed by this time, independently by Albert Grünbaum at the Hygienic Institute in Vienna and Fernand Widal in Paris; what is now referred to as the serological Widal test involved the ability of serum from an animal, human or otherwise, to agglutinate typhoid bacilli. As was the situation described above in determining who should have priority in the identification of the etiological agent of typhoid, priority in development of the serological test was also in dispute. Widal would appear to have had priority.

> On June 26th last [1896] I brought before the Medical Society of the Paris Hospitals a new method—to which I gave the term "sero-diagnosis"—by which typhoid fever could be recognized almost instantaneously by simply observing microscopically how the serum of a patient acted on a culture of Eberth's bacillus. I described several processes to establish the sero-diagnosis, some being slow and others rapid. Among the slow processes one consisted in adding a few drops of serum to a culture of Eberth's bacillus in bouillon already cloudy, and observing a few hours later whether the bouillon became clear again and if, at the same time, a precipitate was formed at the bottom of the tube. The other process consisted in adding to a tube of fresh bouillon a few drops of serum, then introducing into it a trace of a culture of Eberth's bacillus, and placing the tube in an oven at a temperature of 37° C, and observing whether after fifteen or twenty-four hours a precipitate had formed at the bottom of the tube, showing under the microscope heaps of microbes, leaving the bouillon almost completely clear.... For ordinary daily practice I prefer the improvised process. I have shown that all that was necessary was to add a drop of serum or even a drop of blood taken from the tip of the finger of the patient to ten drops of a young culture in bouillon of typhoid fever bacilli, and to see almost immediately under the microscope of "heaps" or agglomerations of bacilli, which often allows an almost instantaneous diagnosis of typhoid fever to be made. I ended my first communication by saying: "Here is a simple and rapid process which can be employed by everyone, necessitating no laboratory material. All that is necessary is to have at one's disposal pure cultures of Eberth's bacillus, a microscope, and a few drops of serum or even only one drop of the blood of a patient."
>
> The agglutination of the microbes by the serum of immunized animals has been under consideration since 1889. As regards the infection of typhoid fever, we have stated previously that Pfeiffer and Kolle had shown that the serum of animals rendered immune against the infection of typhoid fever as well as the serum of convalescents from typhoid fever, injected into the peritoneal cavity of a guinea-pig at the same time as a culture of Eberth's bacillus, possessed the property of rapidly immobilizing, of agglutinating, and deforming the microbe present in the serous cavity. We also mentioned that [Herbert] Durham and [Max von] Grüber had previously shown that the serum of animals rendered immune against experimental typhoidal infection had the power of immobilizing and uniting in heaps or masses *in vitro* Eberth's bacilli in suspension in a liquid.[51]

Grüber and Durham tested only animal serum. They did not attempt to actually determine whether this would be applicable to human diagnosis.

At approximately the same time as Widal presented his work, Albert Grünbaum independently described a similar process which could be used to diagnose typhoid fever.

> If a drop of an emulsion of a motile pathogenic organism is mixed with the drop of the serum of an animal immunized against this particular bacillus the microorganisms collect together in "clumps" and lose their mobility. The importance of this reaction, which had been seen by others, was first recognized and studied by Grüber and Durham. They threw out the suggestion that it might be made of diagnostic value in the sense that bacteria cultivated from the stools of cholera or typhoid fever patients might be identified by its aid. But it appeared likely on *a priori* grounds that the "agglutinins" found in the serum of immunized animals would also be formed in the human body during an attack of enteric fever or cholera (it being already known that "protective" substances, at any rate, are so formed); and at Professor Grüber's suggestion I undertook the investigation of this point as part of the question whether immunity and protection are not (in certain diseases) dependent on and proportional to the agglutinins present. The serum of normal guinea-pigs has rarely any pronounced agglutinative action, but it does not follow that the same would hold good for man, and an examination of some thirty cases of normal and diseased individuals ... showed the presence of agglutinins in persons not suffering and not having suffered from enteric fever (or cholera). Although anticipating matters, it may be here stated that hitherto *it is only in cases of enteric fever that the serum shows a distinct agglutinative action within thirty minutes when diluted sixteen times* [italics in original], and hence this reaction can be used as a diagnostic sign. If the reaction occur in still greater dilution its diagnostic value is correspondingly increased.... [Grünbaum completed his report with a comment about Widal's observations.] Whilst this work was in progress a short communication by Widal applying a similar method macroscopically has been published in the *Semaine Medicale*. It requires, however, a larger quantity of blood and a longer time, and apparently, from the rather meager description, a final resort to the microscope; and in one case, at any rate, I was unable to detect the reaction.[52]

Widal, of course, could not avoid responding to Grünbaum's final comments.

> Up to June 26th, 1896, the date of my first communication, the phenomenon of agglutination had been considered as a "reaction of immunity" appearing only in immunized animals. I was the first to show that it was indeed a "reaction of infection," that it appeared in man during the first days of the disease, and I thus arrived at the conception of sero-diagnosis and its applications. A few days after my first communication, confirmative facts were presented by different clinicians and today some hundreds of observations of positive sero-diagnosis have already been published in Europe and America. Up to the present day my improvised process has almost always been preferred by experimenters. It is the same process which M. [Charles] Achard and M. [Raoul] Bensaude have employed for the sero-diagnosis of Asiatic cholera. I was, therefore, very much

surprised to find an article on the Agglutinative Action of the Human Serum for the Diagnosis of Enteric Fever, published in *The Lancet* of September 19 last—that is, nearly three months after my first publication—by Dr. Grünbaum, who proposes a much more complicated process than my extemporaneous method, and only adds at the end of his article, as history of sero-diagnosis, the following: "Whilst this work was in progress a short communication by Widal applying a similar method macroscopically has been published in the *Semaine Medicale*. It requires, however, a large quantity of blood and a longer time and, apparently from the rather meagre description, a final resort to the microscope." If for a purely technical question Dr. Grünbaum, instead of being contented with an incomplete analysis given by a paper, had referred to the official text of my researches or to the *Presse Medicale*, which reproduced *in extensor* most of my works, he would have seen that I did not publish a *short communication* [italics in original] on the sero-diagnosis, but that since June 26th last I have published a series of articles on this subject, and that my favorite method is based only on microscopical examination, is extemporaneous and only necessitates a minimum quantity of blood. Dr. Grünbaum would have avoided an error which I desire to correct.[53]

Widal further included as a note the more recent application of his diagnostic technique.

I may add that in collaboration with M. Sicard I brought forward on September 29th at the Paris Academy of Medicine, and on October 9th and 16th at the Medical Society of the Paris Hospitals, new researches on sero-diagnosis and experimental and chemical researches on the agglutinative phenomenon which are too long to be considered here. I may also add that at present we have practiced sero-diagnosis in forty-five serious as well as mild cases of typhoid fever during its acute stage, and that we have searched for, without finding it, the agglutinative reaction of the blood of more than 200 individuals in good health or suffering from other diseases, by mixing the serum and culture of Eberth's bacillus in the proportions which I have indicated. The different authors who have controlled by researches have arrived at the same results."[54]

In summarizing the products of research described above, it is clear that mechanisms already were in existence by 1898 to not only differentiate typhomalaria, as the misdiagnosis was called, from typhoid fever, but also both a proper method for diagnosis as well as prevention of the problem in the first place. One can only conclude that it should not have been necessary to create the Typhoid Commission in order to understand the outbreak of the disease in southern army camps.

Chapter 7

The Interim, 1898–1916

Vaughan was discharged from the army in June 1899. Despite the deaths of the other two members of the commission, Shakespeare in 1900 and Reed two years later, he was able to complete and submit an abstract of the report by June 1900, and produce a complete report in 1904. By this time Vaughan had returned to his position as dean of the medical school at Michigan. Among the changes which he instituted for future admissions was a requirement that beginning in 1909, all students had to have completed the equivalent of two years college work. Course work for the first three years of the medical curriculum would include gross anatomy, physiology, bacteriology, physiological chemistry, pathology and pharmacology; much of this work would include laboratory studies as well. Surgery and other specialized activities would constitute large portions of both the third and fourth year of the medical program.[1] These changes brought the medical curriculum at the university more in line with those of European schools such as those in England, France or Germany.

The change in the medical school curriculum shepherded by Vaughan was only a small portion of the changes which were taking place nationally in the training of future physicians. As a member of the American Medical Association and serving as chairman of the Reference Committee on Medical Education, which was originally set up in 1902, he recommended the creation of the Council on Medical Education, a committee that became instrumental in establishing the reforms which had been in place at Michigan and which would take effect across the country.[2] Members of the council included Drs. Arthur Dean Bevan, professor of surgery at Rush Medical College in Chicago as chairman, J.A. Witherspoon, professor of medicine at Vanderbilt University School of Medicine, William T. Councilman, professor of pathology at Harvard Medical School, Charles H. Frazier, professor of surgery at the University of Pennsylvania in Philadelphia, and of course Vaughan himself.[3]

Changes in the medical programs had as much to do with political interests as that of the scientific intent. The changes in the licensing of medical schools which began in the 1890s reflected the growing interest in research as the national organization began to control the state and local boards. It was no coincidence that the five original members of the council were all involved in research as well as in medical education itself; all had either attended European schools, or at least were strongly influenced by them; Vaughan as noted earlier had had the opportunity to spend time working in the laboratory of Robert Koch.

The council went about its business by first designing the state licensing board examinations, with questions emphasizing those areas of medicine most studied in so-called "scientific medical schools." Emphasis was placed on the basic sciences, while areas of therapeutics and *material medica*—homeopathy—were de-emphasized. In this way schools of homeopathy were placed at a significant disadvantage.[4]

The changes instituted by the council evolved in two steps. The initial, albeit temporary, requirement was that the prospective student would first finish a four-year high school preliminary education. If accepted into a medical school, the student would complete a four-year program similar to that outlined by Vaughan at the University of Michigan Medical School, and finally pass the examination administered by the state licensing board. The "Ideal Standard," proposed by the council in 1905, changed the four-year medical program into one of five years, including two years of clinical training and a sixth year serving as an intern. As outlined in the "ideal," the typical medical student educated in the United States, not including the four years in high school, would spend two years in a university, four years in medical school, and another year as an intern.[5] The changes as they were instituted would become increasingly familiar to the medical student of the twenty-first century.

The council began by examining the quality of the existing medical schools, initially ranking them in classifications reflecting the proportion of graduates capable of passing state board examinations. Two subcommittees were established, one which would report on admissions requirements, consisting of council members Vaughan, Bevan and Witherspoon, and a second subcommittee, consisting of Bevan, Councilman and Frazier, which would prepare a report discussing the curricula at the various medical schools.[6]

The report was completed by 1907. Fifty-four of the poorest ranking schools fell into only five, largely Midwestern, states. In total, one hundred and sixty schools were visited and subsequently placed into one of three classes; ranking was based in part upon graduates passing the state examinations, and including (subjective) grading of hospital, laboratory and library facilities. A total grading of 100 points was possible. Schools placed in Class A scored 70 or higher, a total of 82 schools. Class B between 50 and 70 and considered doubtful, totaled 46 schools. Class C, marked below 50 and considered unacceptable, comprised 32 schools. If a school was considered unacceptable it was forced to either raise its standards or go out of business. State licensing boards were forced to raise their standards as well; simply granting a license to practice medicine to anyone who graduated from a so-called medical program would no longer be the norm.

The result was a significant reduction in the number of medical schools, reaching a level of 126 by the time of the next inspection in 1910. (By 1920 the total number of American medical schools was reduced to 85 and by 1927 the number had fallen to 80.)[7] The University of Michigan Medical School, thanks largely to the reforms of the previous decades, many of which had been instituted under the tenure of Vaughan, placed in the top ranking. Only the medical schools at Harvard and Johns Hopkins required a college degree prior to acceptance into their respective programs.

Vaughan summarized the findings of the council:

> At the lowest extreme [Class C] were a few institutions which were actually selling diplomas; in a large number only didactic, lecture or recitation courses were given, sometimes by the one teacher who constituted the entire faculty. Most of the schools had no laboratories save an old time dissecting room and occasionally an excuse for a chemical laboratory. One institution was found which turned out one hundred and five graduates in 1905, without having completed any laboratory work, not even dissecting, nor had they had the opportunity to see a single patient in either a dispensary or a hospital. Less than half the colleges had affiliations with either dispensaries or hospitals in which patients could be utilized for clinical instruction, and the schools were few, indeed, in which students had the opportunity to study patients in small clinics at the bedside or as clinical clerks.[8]

To nobody's surprise, the result of the report was a significant "furore" [sic], much of it directed at Vaughan and the other members of the council. In response to the uproar, much of it of course political, the council requested that the Carnegie Foundation carry out a similar survey, with the primary difference being the person in charge would not be a physician. The foundation chose educator Abraham Flexner—a member of the Carnegie Foundation, and though not personally a physician, he was the brother of medical researcher Dr. Simon Flexner—to prepare the report. The findings of the Flexner Report largely mirrored those of the council: "There has been an enormous over-production of uneducated and ill-trained medical practitioners ... due in the main to the existence of a very large number of commercial schools, sustained in many cases by advertising methods through which a mass of unprepared youth is drawn ... into the study of medicine.... Colleges and universities have in large measure failed in the past twenty-five years to appreciate the great advance in medical education and the increased cost of teaching it along modern lines."[9]

The council was vindicated. However, Vaughan went on to warn his readers some years later, ca. 1920s, that the two reports did not bring to an end what was clearly a corrupt system. In 1925 "diploma mills" were found in two states, Missouri and Connecticut, in which members of the state boards of licensing had been subjected to bribery. There were likely additional diploma mills elsewhere, in other states, which were not located. Within the recent memory of Vaughan as he was writing his autobiography, newspapers carried advertisements of so-called medicines, "and some of the manufacturers of these worthless and often harmful nostrums accumulated great wealth and influence."[10]

One may speculate that in this arena of medicine some things rarely change. Anybody watching television or reading various forms of written media is subjected to advertisements for similar worthless nostrums such as colon cleansers, hair restorers or various means to remove toxins from the body. The disclaimer—often in the same color as the background and rarely on the screen for longer than a brief moment—always indicates such treatments are not for the curing of specific diseases and do not have FDA approval.[11]

Vaughan continued as a member of the council for much of the remainder of his professional career; in 1913 he was elected president of the American Medical Association, serving until 1915.

The council's success in eliminating at least a large majority of such diploma mills reflected not only the advances in medical knowledge in the previous decades, but changes in the political climate as well. The Flexner Report did not represent the first attempt at regulation of the profession, merely the most successful. Among the most egregious of these "diplomas for sale" was that of a medical college in Boston.

> In 1880, Rufus King Noyes, an anti-vaccinationist of some notoriety in Lynn, Mass., obtained a charter of incorporation for the Bellevue Medical College of Massachusetts—a colorable imitation of the title of the Bellevue Medical College of New York. Two years later a "diploma" of this concern, signed Rufus King Noyes M.D., President, was presented to Dr. [John] Rauch, then Secretary of the Illinois State Board of Health, and a license to practice medicine was demanded thereon. An investigation of the Boston "Bellevue" was at once set on foot, during which letters written by a young journalist resulted in an agreement by Noyes to furnish for $150 tickets showing attendance upon two courses of

lectures, and a "diploma" of graduation from his "college" to a man whom he had never seen; who had never been in Boston in his life; who only claimed to "hav bin Redin medesin [sic] about a year"; but who, having "ben tending on Sick purty Near all [his] life," thought he had a "Purty good Idee about the bizness"; [sic] and who furnished a thesis on "Vacinnatioun," in which he professed to be ready to make 'a Strong kick on the Part of our noble Proffession [sic] against the Inseartion into the Pure Blood and Vitle fluid of our Inosent offspring of that Diseas of the animals cow-pox."

This correspondence was begun in October, 1882, and on November 2, 1882, Noyes wrote: "Dear Sir:—You, as a candidate for graduation, have been favorably considered by the faculty, and your thesis has been examined by the Professor and found to be acceptable.... You are correct on the Vaccination question, and I am confident you will meet with continued success. Your diploma will be sent C.O.D. one week from the date of this letter. It will be securely packed in a pasteboard box. Your bill for diploma and two tickets is $150."

On receipt of the "diploma" it was promptly turned over to the United States authorities, together with the correspondence, as the basis of a prosecution for illegal and fraudulent use of the United States mails—that being, as it was advised, the only ground upon which a charge could be made. During the trial Noyes and his codefendents pleaded that they were legally incorporated and empowered by the laws of Massachusetts to issue diplomas and award degrees without any restriction as to course of study or professional attainments. The United States Commissioner held the plea to be valid and dismissed the case, saying: "The law makes the faculty of the college the sole judges of the eligibility of applicants for diplomas. There is no legal restriction, no legal requirements. If the faculty choose to issue degrees to incompetent persons, the laws of Massachusetts authorize it."[12]

While Noyes clearly established his "school" for monetary reasons, he actually was a legitimate physician, having received his degree from Dartmouth in 1875. With the publication of the Flexner Report, "schools" such as that established by Noyes were no longer protected by legal technicalities.

Vaughan and his colleague Frederick Novy spent much of their time during these years promoting public health programs around the state. He and Novy each strongly emphasized to their audiences the importance of public health inspections of water supplies. In addition to his regular teaching duties, Vaughan also presented a yearly lecture on sex hygiene to the university students—"Attendance was required for men, but hundreds of women voluntarily went to lectures given especially for them."[13] The purpose of these lectures was the attempt to provide students with accurate, albeit contemporary, information concerning sexual practices in an informal manner. Stu-

Victor Vaughan in Laboratory (Bentley Historical Library, University of Michigan [Victor Vaughan file]).

dents were encouraged to freely ask questions. It was not unusual for Vaughan to include an element of humor in his responses. One anecdote provided by Davenport in his history of the medical program at Michigan related to a former student who was unable to break a habit of masturbation. Vaughan's recommendation was to "marry some nice girl and forget about it."[14]

During these years Vaughan also received recognition from the medical school faculty on the occasion of the twenty-fifth anniversary of having received his doctorate. On June 18, 1903, University President James Angell presented a bound volume of thirty-four papers, "Contributions to Medical Research Dedicated to Victor Clarence Vaughan," written by his students and colleagues.[15] Angell began his presentation by briefly outlining the changes which had taken place in the medical curriculum during the quarter century in which Vaughan had developed from being a promising student to his role as dean of the medical school. When Vaughan entered the school, instruction consisted nearly entirely of lectures by the faculty, and it was not unusual for students to listen to the same material repeatedly. Among the most significant changes instituted by Vaughan was the incorporation of significant amounts of laboratory work, providing the students with an hands-on experience in medical subjects.[16] In acknowledging the award and thanking the faculty and president for the honor, Vaughan also pointed out the importance of research in maintaining a high quality medical program.

Vaughan himself maintained an active research program throughout his professional career. A significant portion of this research addressed the roles of bacterial toxins and their relationship to disease. As Davenport pointed out, a compilation of Vaughan's numerous books, reviews and presentations on the topic produces a highly repetitive series of publications.[17] Any analysis of Vaughan's work on bacterial toxins illustrates the confusion which was a direct result of the understanding of the nature of toxins inherent to the time. While bacteria had been observed microscopically for hundreds of years, and their role as etiological agents of infectious disease had been confirmed largely during the 1880s and '90s, the nature, and certainly the structure, of the organisms had yet to be defined. In 1884, the Danish physician Hans Christian Gram had observed that bacteria could be subjected to differential staining. Gram's original procedure consisted of initially staining the schizomycetes (bacteria) with an "aniline-gentian violet solution followed by immersion in a solution of potassium iodide after a rinse in absolute alcohol, which resulted in a decolorization of certain species of bacteria."[18] Gram observed that some bacteria, the "cocci of croupus pneumonia, cocci of pyemia [streptococci] ... anthrax bacilli" retained the blue-violet color, while other bacteria such as those associated with typhoid (*Salmonella*) were decolorized and appeared reddish.[19] What is now called the Gram stain is considered among the most important methods used for the identification of bacteria. Gram's purpose was largely to develop a means to stain bacteria against a background of tissue, and the significance in using the procedure as a means of differentiation was lost on him. Gram was hardly unique in overlooking this possible application of his staining procedure, with the possible exception of German physician Carl Flugge, who noted in his textbook of micro-organisms: "The method of Gram is mainly useful for the differential staining of bacteria in tissues and for the diagnostic differentiation of species."[20]

Vaughan believed that

> we have been led by observation of many facts concerning the pathogenic bacteria and their action in the various infectious diseases … to formulate a theory which we believe to be in some respects new, and which may be stated as follows: The specific poisons of pathogenic bacteria are formed within the bacterial cell and constitute part of the organism itself … the heightened virulence is due to an increase in the amount of toxin formed within the cell or to a greater permeability of the cell wall.… That the cell walls differ in the readiness with which their toxin contents diffuse out of them seems evident from the way in which diphtheria and tetanus bacilli induce intoxications while their growth is confined to a very limited area, as compared with the relative innocuousness of the colon bacillus growing abundantly in the intestines, and at the same time containing a very active toxin.[21]

Unknown at that time (ca. 1901) was the difference in the nature of the toxins produced by the respective bacteria. Diphtheria and tetanus each produce a secreted toxin, the former first demonstrated by the German physician Arthur Nicolaier in 1884, and the latter by Emil von Behring (among others) later that decade. What was not known by Vaughan was the nature of the toxin produced by the colon bacillus (*Bacillus coli communis/Escherichia coli*) or other such Gram negative bacteria, the active component of which was actually part of the cell structure. The logical first step in addressing the problem was to isolate the toxin and determine its chemical makeup. "When bacterial cell substance … is heated in a flask with a reflux condenser, with from 15 to 25 times its weight of a two percent solution of sodium hydroxide in absolute alcohol, the cell molecule is split into toxic and non-toxic groups. The poisonous portion is now in solution in the alcohol and in the case of b. coli [*Bacterium coli*] and b. typhosus [*Bacterium typhosus*] it constitutes by weight about one-third of the cell substance."[22]

Vaughan went on to describe the purification procedure. The toxic portion, which Vaughan designated as the "crude bacterial poison," was soluble in alcohol.[23] The partially purified extract was shown to be extremely toxic in a variety of test animals. Vaughan was unsure of the chemical nature of the poison, suspecting that it may have been composed in part of protein, an assertion based upon several chemical tests and experiments to which he referred as being carried out in his laboratory which demonstrated similarities with egg albumin and peptone; however, the solubility in alcohol was more indicative of the toxin being a lipid. Vaughan even referred to the isolated substance as an endotoxin.[24]

It would not have been a stretch in Vaughan's imagination to suspect the bacterial poison was a component of the membrane portion of the outer portion of the (Gram negative) cell. In 1892 Richard Pfeiffer, an associate of Robert Koch, while studying the toxic activity of *Vibrio cholerae*, had observed that guinea pigs which had been immunized against the cholera bacillus could be killed if exposed to either living or dead cholera bacilli. Pfeiffer's (correct) interpretation was that following lysis or breakdown of the bacteria, a poisonous substance was released from the dead cells. The substance was heat stable, suggesting it was not composed of protein. The same substance was identified in a variety of Gram negative bacteria, and was subsequently termed, either by Pfeiffer himself or one of his colleagues, as being an endotoxin, a component of the cell structure itself.[25] Interestingly enough, while Pfeiffer is properly given credit for the discovery of what is now known as endotoxin, in an argument one might find only among scientists, it is unclear whether Pfeiffer himself used the term in his work prior to 1900. It appears Pfeiffer's first use of the word endotoxin was in lectures he presented in 1903 or 1904.[26]

In fact, Vaughan during this time appears to have drawn an incorrect conclusion regarding the nature of the toxin he was investigating. A contemporary of Vaughan's, Hans Zinsser, physician and subsequent author of numerous popular literatures in the medical field, later (1922) compared Vaughan's observations and conclusions with those of Pfeiffer's.

> The majority of pathogenic bacteria do not ... produce *true toxins* or *exotoxins*. Cultures of cholera spirilla, plague bacilli, and of many other bacteria [all Gram negatives] do not yield toxic filtrates until the cultures have been allowed to stand for prolonged periods during which extraction and possibly autolysis have occurred. In these cases, moreover, definite toxic properties can be demonstrated in the dead cell bodies or in extracts prepared by various methods. In no case, however, is the injection of these "endotoxins" followed by the production of antitoxins. It was very natural to suppose that in microorganisms of this class the toxic principle might be present in the form of a preformed intracellular poison which could be extracted or became free as cell-death occurred and disintegration ensued.
>
> It was assumed that, when bacteria entered the animal body and were destroyed by the action of the serum or cells, these endotoxins were liberated and poisoning resulted. The very protective action of the serum, which prevented the extension of the infectious invasion, by limiting bacterial growth,

was thus looked upon as the agency by which the endotoxins, toxalbumins, were set free.... [Zinsser was correct in his representation of the toxic substance being liberated upon the death of the bacterium. The substance, however, was not a form of albumin.] Experiments in which it was shown that large doses of bacteria injected into immunized animals were violently toxic and more rapidly fatal than corresponding amounts injected into normal animals, were taken to mean that in the immune animals a more powerfully cell-destroying property of the serum led to a more rapid liberation of the endotoxins.

This was the conception of Pfeiffer.... Its essential features consisted of the assumption that the poisons were preformed and were contained within the cell body as such, and that they were specific for each micro-organism, determining to a certain extent its pathogenic properties. Thus typhoid endotoxin, cholera endotoxin, or dysentery endotoxin [all are Gram negative cells] was supposed each to contain its own particular pharmacological properties by which the clinical manifestations of the respective diseases were partially determined. [This did not rule out the fact that exotoxins might also be involved with pathogenic properties, as with cholera exotoxin.]

It is chiefly the work of Vaughan which has begun to throw doubt upon Pfeiffer's original views, in that Vaughan has shown that all proteins, bacterial or otherwise, would yield ... toxic split products which possessed many of the pharmacological properties of the so-called endotoxins. In fact, Vaughan succeeded in producing, in animals, fever and other symptoms which are generally associated with infection, merely by injecting into them graded quantities of his toxic split products.[27]

Further evidence for the nature and source of the toxic substance was also overlooked by Vaughan when he tested the relative toxicity of his extracts from a variety of bacteria when injected into test animals: guinea pigs. For example, measuring toxicity as a function of the animal's body weight, Vaughan observed that extracts from *Bacillus prodigiosus* (*Serratia marcescens*) were nearly 50 times more toxic than that from *Bacillus anthracis*, the etiological agent of anthrax. Extracts from *Bacillus coli* (*Escherichia coli*) were nearly 40 times more toxic. In fact, extracts from Gram negative bacteria were uniformly more toxic than that from Gram positives. Two additional characteristics also argued against the protein nature of the toxins: toxins were within the cell structure and were not secreted into the medium (unlike the situation with anthrax or diphtheria toxins), and high temperatures did not destroy the toxic activity, suggesting a lipid or lipopolysaccharide rather than a protein content.[28]

In his defense, not only the chemical difference between what are now recognized as endotoxins and exotoxins was unclear at the time—the first decade of the twentieth century—the actual chemistry of the bacterial cell was subject to debate. Vaughan and his students spent considerable time during the next decade attempting to completely purify and identify the toxic substance, a venture which met with only limited success. It was only during the 1930s, some years after Vaughan's death, that French bacteriologist Andre Boivin and his Romanian colleague Lydia Mesrobeanu confirmed the lipopolysaccharide nature of the endotoxin ("antigène glycido-lipidique").[29]

In 1902 Vaughan and Novy published the fourth edition of *Ptomaines and Leucomaines*, now titled *Cellular Toxins; or, the Chemical Factors in the Causation of Disease*.[30] Vaughan and Novy's views on the etiological role of bacteria and their toxins in disease had evolved by this time, incorporating the latest scientific research and knowledge on the subject. This especially applied to their understanding of bacterial toxins (ptomaines) as well as the nascent field of immunology which now emphasized the research being carried out in Germany. The etiological causes of disease (as originally defined in the 3rd edition [1896]), were now placed in seven different categories: (1) bacterial, (2) fungous, (3) protozoal, (4) animal parasitic, in which the life cycle of the organism includes a parasitic stage, (5) intoxications, poisonings which result from the presence of agents which originate outside the body, (6) traumatic, resulting from external violence, and (7) autogenous diseases. The authors acknowledged that categories may overlap; for example, external trauma may allow for infection by a bacterial agent.

Vaughan and Novy again focused on anthrax as the prototype disease resulting from elaboration of a chemical poison, describing as well that the pathology associated with other diseases such as diphtheria, cholera or tetanus likewise resulted from elaboration of toxic compounds. Indeed, their definition of an infectious disease was based upon this premise: "An infectious disease arises when a specific, pathogenic microorganism, having gained admittance to the body and having found conditions favorable, grows and multiplies, and in so doing elaborates a chemical poison which induces its characteristic effects."[31]

The elaboration of a toxin by the pathogen continued to remain as the central theme in the theory behind pathogenic mechanisms. Much as they had in earlier editions of *Cellular Toxins*, often using the same examples, Vaughan and Novy reached this conclusion through playing a form of devil's advocate: if other pathogenic mechanisms were involved, one would expect to find certain experimental results. In the absence of those results, chemical toxicity remains the logical conclusion.

Many of these examples as they applied to anthrax had been previously addressed in the first edition of *Ptomaines and Leucomaines*. German pathologist Otto Bollinger had theorized that since the bacilli which are the agents of apoplectiform anthrax, perhaps the most dramatic form of the disease, in which seemingly healthy animals suddenly die, are aerobic in that the host (cattle or sheep) is asphyxiated; red blood cells are deprived of oxygen. Some symptoms of anthrax seemed to suggest such a mechanism: cyanosis and dyspnea. Post-mortem examinations also seemed to support the theory since the dead animals often had distended veins as well as organs which appeared cyanotic. But as Vaughan had previously pointed out, for Bollinger's theory to be correct, one would expect to find large quantities of bacteria in the blood. This was rarely the case. Vaughan also pointed out that Von Nencki from the St. Petersburg Institute for Experimental Medicine in Russia had directly tested the ability of animals sick with anthrax to carry out a form of physiological oxidation, a process which would depend upon the presence of sufficient oxygen. Von Nencki found the oxidation process to be normal, indicating death is not the result of oxygen deprivation. Since the anthrax bacillus is aerobic, requiring oxygen to survive, and many other pathogens are anaerobic, either inhibited by oxygen or at least not requiring it in order to grow, "the theory that germs destroy life by depriving the blood of its oxygen has been found not to be true for anthrax, and if not true for anthrax, certainly it cannot be for any other known disease."[32] As pointed out (and quoted) in an earlier chapter, Vaughan and Novy's interpretation was the same and their conclusion had not changed.

If a chemical poison is the basis for the pathology associated with infection as Vaughan and Novy concluded, how did they theorize its production? While they did not rule out the possibility that cleavage products of cell proteids could be involved since "it is well established that certain powerful poisons originate in this way," there were significant problems associated with this theory. "Among the bacterial split products formed either in artificial culture media or in the body, there is not found one which, on account of its intensity of action or from the nature of the symptoms which it produces, can be regarded as the specific cause of any one of the infectious diseases. Moreover, it has been shown that some of the most virulent germs, as for instance, the bacillus of tetanus [isolated in 1889 by Japanese bacteriologist Shibasaburo Kitasato while working with Robert Koch], will grow and retain their virulence in artificial cultures made up principally of inorganic substances and containing only minute quantities of organic bodies of such simple construction that it must be admitted that the specific toxins of these microorganisms cannot result from their cleavage action."[33]

The authors still concluded, "Poisons may be produced by the cellular activity of bacteria much in the same way as morphin [sic] is formed in the poppy."[34] Vaughan and Novy had used the same analogy (and wording) in earlier editions of the book, even as recently as 1896.[35] But much had been discovered during the interim, including the demonstration of toxins associated with other pathogens. Explanations were found as to the likely source of the toxin: they were actively synthesized by bacteria.

It is now generally believed that most, if not all, of the pathogenic microorganisms consist of cell walls containing cell protoplasm, and that the specific toxin is a constituent of the protoplasm, and that its formation is one of the vital phenomena manifested by the organism in its processes of growth and multiplication. In some species the cell wall is not easily permeable and the toxin is found only within the cell; while in other species the toxin formed within the cell readily passes through the wall and diffuses through the culture media in artificial growths, or through the tissues, when the germ is multiplying in the animal body. In at least some species the formation of a toxin is not a phenomenon which invariably accompanies growth and multiplication. This is shown to be true by the frequently observed fact that a highly virulent germ may under certain conditions wholly lose its toxicity while it continues to vegetate most luxuriantly. It seems to be evident that certain conditions of growth, as, for instance, the nature of the medium, the temperature, the supply of oxygen, and the presence or absence of certain chemical agents, determine the amount of specific toxin formed within the cell. Most pathogenic germs find the conditions suitable for the elaboration of their poisons at their optimum in the animal body, and for this reason their virulence is increased by passing successively through a series of animals.[36]

Exceptions to an increase in virulence also existed, as the authors pointed out. More importantly, the authors had incorporated into their ideas some of the findings from the previous decades as to the nature of bacterial toxins (discussed in a previous chapter). Toxins may constitute a portion of the components of the cell wall, what we refer to as endotoxin, or may be secreted by the cell: exotoxins.

This theory also proposed that toxins could either be produced and act locally, or disseminated systemically. Vaughan and Novy attempted to provide examples for each.

> In the systemic infectious disease, such as anthrax, typhoid fever and cholera, the specific poison is undoubtedly taken into the general circulation, and may reach and influence every part of the body. In the local infectious diseases, such as gonorrhea and infectious ophthalmia [likely what is sometimes called pinkeye], the first action of the poison seems to be confined to the place of its formation; although even in these, when of a special virulent type, the effects may extend to the general health, or the poison may act strongly on some distant part of the body. It is probably true that in many of the infectious diseases the chemical poison has both a local and a systemic action; thus, it is by no means certain that the ulceration of typhoid fever is due directly to the living bacillus, for it is now an established fact that this disease may exist, run a typical course, and end in death, without anatomical changes in the intestine. In diphtheria and tetanus the toxin formed within the bacterial cells readily diffuses through the cell walls and enters the circulation, while the organism itself is confined to relatively a small area and may not be found in the blood at all. Such diseases are properly called bacterial intoxications. In some other infectious diseases, such as anthrax and one form of the plague, the germ itself may be distributed by the

blood and lymph to every part of the body; these diseases are designated as septicemias.[37]

It is now understand that pathological effects found in diphtheria, anthrax, cholera are associated with toxin production, though in the case of cholera primarily confined to the intestine; Vaughan was correct in this representation. Until recently it remained largely a mystery how the typhoid bacillus, *Salmonella typhi*, produced a wide variety of symptoms in different hosts. In 2013 a toxin was identified and at least some of the questions explained.[38] While Vaughan included the plague bacillus among those bacteria which produce a chemical poison, the products produced by the organism, now known as *Yersinia pestis*, are associated primarily with evasion and survival in the face of the host's immune system rather than being strictly an intoxication.

A better understanding of the immune response to infection was among the most important advances in medical science in the years following publication of the 3rd edition of Vaughan and Novy's treatise. In their studies of the toxin produced by the diphtheria bacillus, German physicians Paul Ehrlich and Emil von Behring had demonstrated both its role in the pathogenicity of the disease as well as the production of soluble proteins in the serum known as antibodies which can neutralize the toxin. The principle of humoral immunity, as it subsequently was termed, became a key factor in the treatment or prevention of major disease.

Vaughan and Novy were well aware of the potential germicidal properties associated with serum following exposure of the host to certain pathogens; some of the early experiments carried out in European laboratories were discussed in the previous edition of the book. Consequently the phenomenon was well known by the early 1900s, even if the underlying mechanism was unclear. As the authors pointed out in a brief review of the topic, as early as 1872, British bacteriologists Timothy Lewis and David Cunningham had demonstrated that bacteria injected directly into the blood of animals are cleared within a few hours. Variations of the same experiment were carried out by additional scientists during the following decade, all demonstrating the same effects of serum on survival of bacteria.[39] In the late 1880s, British bacteriologist George Nuttall demonstrated that when defibrinated blood, blood which had been allowed to clot, was mixed with a variety of pathogenic bacteria, including the anthrax bacterium, that the organisms were killed.[40]

> The importance of this discovery was at once recognized, and this property of the blood has been regarded as one of the chief defenses of the human body

against the invasion of pathogenic bacteria. Much inaccuracy has arisen, however, in the general conception of this property of the serum from a failure that the action of human and of animal serum may be very different upon the same bacterium, and, furthermore, that all kinds of bacteria are by no means equally affected by human serum, some being promptly destroyed by it, while others are hardly affected.... The source and chemical nature of the germicidal substance in blood, the so-called "alexin," [now known as complement] is still an open question. [German bacteriologists Hans] Buchner, [Rudolph] Emmerich, and many other scientists believe it is a serum-albumin or albuminate, Vaughan considers it a nuclein and Ogata [Masanori] a ferment. The white corpuscles of the blood seem to be intimately concerned in its production and experiments ... show that it is a product of the life and metabolic activity of the polynuclear leucocytes, not a result of their death and degeneration.... The changes in the germicidal properties of the blood serum which occur during and after an acute infectious process are important, though but little understood at present.[41]

The description by the author of the journal article, Franklin Warren White, is that the germicidal substance is a component of the blood serum, and is either the product of white blood cells, or at least is involved in some form of interaction with those cells. Hans Buchner was among the first to describe an antibacterial substance he called alexin—a protective substance—in serum which exhibited germicidal properties. Paul Ehrlich later termed the substance "complement."

There was dispute over whether this was the result of phagocytosis by white blood cells (unlikely in defibrinated blood), the French school of thought following the work of Elie Metchnikoff at the Pasteur Institute in Paris, or from undefined substances soluble in blood, the German school of thought. The characteristics of the substance, or at least some of its antibacterial properties, gradually became clearer through the work of Buchner and his students during the 1890s. As summarized by Vaughan:

(1) The germicidal action of the blood is not due to phagocytes, because it is not influenced by the alternate freezing and thawing of the blood, by which the leucocytes are destroyed. (2) The germicidal properties of the cell-free serum must be due to its soluble constituents. (3) Neither neutralization of the serum, nor the addition of pepsin, nor the removal of carbon dioxide gas, nor treatment with oxygen, has any effect upon the germicidal properties of the blood.
(4) Dialysis of the serum against water destroys its activity, while dialysis against 0.75 percent salt solution does not. In the diffusate there is no germicidal substance. The loss by dialysis with water must be due to the withdrawal of the inorganic salts of the serum. (5) The same is shown to be the case when serum is diluted with water and when it is diluted with salt solution; in the former instance the germicidal action being destroyed, while in the latter it is not.
(6) The inorganic salts have in and of themselves no germicidal action. They are active only in so far as they affect the normal properties of the albu-

minates [proteins] of the serum. The germicidal properties of the serum reside in its albumin constituents. (7) The difference in the effects of active serum and that which has been heated to 55° is due to the altered condition of the albuminate. The difference may possibly be a chemical one (due to changes within the molecule) or it may be due to alterations in mycelial structure. The albuminous bodies act upon the bacteria only when the former are in an active state.[42]

The understanding of these properties of alexin/complement observed by Buchner and outlined by Vaughan only came about later. The complement pathway consists of a series of proteins, many of which are inactive enzymes until the pathway is activated. A variety of co-factors are necessary for several of the steps, hence the effects of dialysis treatment. Vaughan was correct in stating that withdrawal of inorganic ions will inactivate the germicidal properties. The complement proteins are also sensitive to heat; incubation at 55° is sufficient to denature ("changes within the molecule") the proteins and eliminate the germicidal activity, a standard procedure today in removing complement from serum.

However, the question addressed at this time by Vaughan and Novy was that of the chemical nature of nuclein, the germicidal substance they had previously described, and whether it might be identical to alexin. Their conclusion was based largely on what could not be the chemical basis of nuclein/alexin. As early as 1893 they had reported that since the activity is not affected by pepsin, a proteolytic enzyme, the material cannot be serum albumin. Several examples of the clinical use of nuclein have been presented in an earlier chapter. However, since it is inactivated by heating at 55°, it must be in the proteid category. Since "the only proteid likely to be in blood serum and which is not destroyed by peptic digestion is nuclein," their conclusion was the material they referred to as nuclein may also be the germicidal substance.[43]

Inactivation at 55° (or higher) was a critical factor in identifying the substance. As Buchner would point out, the activity of alexin is inactivated at that temperature, while that of nuclein (nucleinic acid) is unaffected at that temperature. Vaughan's response to this observation was that the germicidal activity of nuclein may be variable. "It is within the range of possibility that there may be a nuclein or nucleinic acid so labile that it loses its germicidal action at the relatively low temperature mentioned above [55°]. It is also possible that while an aqueous solution of nucleinic acid may retain its germicidal properties at 100°, when mixed with the constituents of blood serum it may be altered at a much lower temperature."[44] Vaughan concluded at this time that the exact nature of the germicidal substances in blood remained unknown.

Vaughan's ideas about immunity to pathogenic organisms or the toxins they produce were also evolving during these years. In many respects the larger concepts of these theories remain unchanged even in more modern times, though of course the molecular mechanisms were unknown in the early 1900s. Vaughan considered immunity to be categorized in three areas: (1) natural immunity, what is now referred to as innate or nonspecific immunity; (2) inherited immunity in which the child acquires immunity through the placental circulation or in colostrum, milk; and (3) acquired immunity, now also referred to as specific immunity, which is acquired through exposure to the organism or toxin naturally or as a result of vaccination. Acquired immunity could also be divided into two forms: active, in which the host is directly exposed to the organism or toxin, and passive immunity, in which the host receives serum from an animal which has been previously immunized. Vaughan used the example of diphtheria to illustrate each. If a child is ill with diphtheria he will respond actively in acquiring immunity against the disease. If the child receives horse serum from an animal which has been immunized with doses of diphtheria toxin, the anti-toxin in the horse serum will neutralize the toxin produced in the child.[45]

Vaughan explained his concept of natural immunity using the analogy of animals being naturally immune to some human diseases. "Among men typhoid fever is one of the grave diseases, causing great morbidity, and increasing to a considerable extent the mortality lists; while among the lower animals this disease does not occur naturally, nor has anyone yet been able to induce it by inoculation with the bacillus. It is true that many of the lower animals are susceptible to the typhoid toxin, but a true typhoidal infection with the bacterium has not been established, although frequently tried, in any of the animals upon which experiments have been made. Leprosy, scarlet fever, yellow fever and measles are other diseases which inflict themselves upon mankind, but to which the lower animals are apparently insusceptible."[46] Though Vaughan did not note this here, the inability to utilize suitable laboratory test animals in studying an etiological agent associated with some diseases has long presented a challenge in applying Koch's Postulates to that agent.[47]

Vaughan attempted to explain natural immunity by suggesting the basis is the inability of the bacterium to multiply within the host. "The bacillus pyocyaneus (likely *Pseudomonas*) is frequently found upon the surface of man's body, especially in the axillary and inguinal regions, and sometimes it occurs in the intestines, but notwithstanding this almost constant proximity of this organism man is but rarely injured by it, yet, nevertheless, this bacillus produces a toxin to which man is susceptible."[48] Vaughan continued with other examples of potentially pathogenic organisms which rarely produce problems in man or other animals with the explanation of an inability to multiply as the cause.

Vaughan largely negated a possible role for alexin in creating this natural immunity. Though alexin was not dismissed outright, Vaughan pointed out that germicidal activity of blood serum against organisms such as the anthrax bacillus did not prevent infection by that agent. Instead he introduced the possible role of phagocytes, cells capable of ingesting and digesting material, as possibly accounting for some forms of natural immunity. Phagocytes had been observed a decade earlier by the Russian bacteriologist Elie Metchnikoff, but his theory of cellular immunity had largely been dismissed by German scientists in favor of the soluble antibacterial substances found in serum.

Vaughan attempted to reconcile these competing theories in his explanation.

> According to Metschnkoff [sic], the most important factor in production of natural immunity is to be found in the phagocytic action of certain cells within the animal body. He divides phagocytes into mobile and fixed elements. Among the former he places both the mono- and polynuclear leucocytes (with the exception of the small lymphocytes, and mast cells of Ehrlich), and the so-called wandering cells [largely what today are known as neutrophils]; while the fixed phagocytes, or macrophages, consist of endothelial cells, the elements of the spleen pulp, and of bone marrow, some connective tissue cells, and possibly certain nerve and muscle cells.... When bacteria are introduced into a naturally immune animal they are seized upon and devoured by either the mobile or fixed phagocytes, or by both. When a microorganism is introduced into a place where no phagocytes are present, as for instance under the skin, into the cornea, or into the anterior chamber of the eye, the mobile phagocytes collect at the point of bacterial invasion, engulf the bacterial cells with the aid of their pseudopodia and then digest them or in some other way deprive them of their capability of harming the body. This method of disposing of foreign substances introduced into the animal body is known as phagocytosis.[49]

Vaughan remained unable to explain the mechanism of killing once phagocytosis had taken place; even Vaughan would have been amazed at the complexity of the process, as has been determined in the age of molecular biology. The logical agent, from his viewpoint, was nuclein, for which he had already demonstrated antibacterial activity. "What this chemical poison is, or whether it is the same in all phagocytes, we cannot at present determine. All phagocytes contain nucleic acids, and the germicidal properties of this have been abundantly demonstrated, but whether or not there may be other and more powerful bactericidal agents in certain phagocytes we have no means at present of knowing." Since it had previously been suggested that disintegration of white blood cells released soluble germicidal contents into the blood—perhaps one of the sources of alexin—Vaughan left open the possibility that this may contribute to the killing ability of the macrophage.[50]

An explanation for acquired immunity remained elusive. Paul Ehrlich had developed a side-chain theory not totally different in principle from the modern elaboration of antibodies, a term Vaughan preferred to avoid.[51] Nevertheless, it was clear that immune serum contained antibacterial specific properties which had not been present in the unimmunized animal.

> When animals are immunized by successive treatments with a microorganism or its products, the blood serum and other fluids obtained from the body acquire either bactericidal or antitoxic properties. In some instances the immunity secured is wholly anti-bacterial, while in others it is antitoxic.... The bactericidal properties possessed by the fluids of the body of the immunized animal may manifest themselves only by an inhibitory action on the growth of the germ, or by partially depriving it of its capability of elaborating toxins. Early in his investigations of this subject, Metschnikoff found that anthrax bacilli grown on the blood serum of sheep immunized to this disease are without effect upon rabbits, but are still possessed of enough vitality to induce fatal anthrax in mice. This indicates that there is something in the blood serum of the immunized sheep which reduces the virulence of the anthrax bacillus.
>
> The question concerning the origin of the antiinfectious substances in the production of artificial [acquired] immunity is one concerning which at present we can do but little more than theorize. The weight of evidence seems to be in favor of the view that by successive injections of the microorganism or its products the leucocytes are stimulated to increased secretion of germicidal substances. Metschnikoff claims [correctly] that it is a general rule that phagocytosis is more pronounced in immunized animals than in those not immunized. He states that when a microorganism is injected into an animal which has been immunized to this germ, the phagocytes of the animal take up the invader more promptly than is done when susceptible animals are inoculated with the same germ. That phagocytosis is more marked in immunized animals is shown by introducing the microorganism in localities ordinarily free from parasites, such as the subcutaneous tissue and the anterior chamber of the eye. When this experiment is made, the phagocytes collect at the point of inoculation very much as they do in an animal possessed of natural immunity, and there they devour the invading organism. The probabilities certainly are that the bactericidal substances found in the serum of immunized animals originate in the phagocytes, whose capability of secreting this substance is heightened by the process of immunization. Undoubtedly it is true that phagolysis also plays a part in increasing the antiinfectious properties of the body juices of immunized animals.[52]

What Metchnikoff had observed, and Vaughan had described, was a process now known as opsonization. In a sense opsonization links the two arms of immunity, cellular and humoral, a complexity unknown to contemporaries of Vaughan. Antibodies, secreted by a class of white cells, lymphocytes, as well as certain components of the complement (alexin) pathway, are capable of binding the surface of bacteria, forming a bridge which in turn will link to receptors on the surface of phagocytes. Since serum from immunized animals would contain such opsonins (antibodies and complement components) the observation that phagocytes in an immunized animal will more readily carry out phagocytosis of the immunizing bacterium was correct.

The contrast between the competing theories of immunity as depicted by German scientists and their counterpart French scientists constituted one of the more significant areas of dispute in understanding acquired immunity during these years. More than scientific knowledge and "truth" represented the underlying basis for the dispute; Germany had humiliated France in the Franco-Prussian War several decades earlier, and the results of that war were still fresh in the minds of their respective citizens. Scientists were hardly immune—if one can insert that ironic phrase—from the same nationalistic feelings. Still, as scientists, each side attempted to support its arguments with experimental observations. The Germans, Koch, Ehrlich, Behring and others, had observed that the presence of soluble protein—alexin—in the serum of immunized animals aided in the killing of pathogenic agents. The French, Metchnikoff and others, had observed a cellular method of killing: phagocytosis. Vaughan, as perhaps a neutral observer, attempted to merge these arguments in developing a view of acquired immunity in which humoral and cellular mechanisms were not mutually exclusive.

Autogenous diseases remained as the final category of illnesses as defined by Vaughan, roughly characterized as diseases which develop within the body in the absence of any external cause such as an infectious agent. As he had explained his theory in earlier editions, Vaughan believed autogenous diseases often have their origins in foods which have undergone only partial digestion. He used the example of peptons [peptone] which may be partially converted into serum albumin following absorption by the cells of the intestine, fats which are converted to glycerin and fatty acids by pancreatic juices. Even if the food is taken in proper quantities and properly digested, if adsorption is not carried out properly the result may be disease.

Vaughan acknowledged that too little scientific information was then available for proper classification of autogenous diseases. Nevertheless, he attempted to delineate characteristics of such diseases as he understood them.

> (1) The digestive organs may but imperfectly perform their function, and the products of their incomplete action may be absorbed and may lead to more or less disturbance in certain organs of the body.... That improperly digested proteids, and in fact certain proteids completely undigested, may be absorbed, is a

well known fact. To what extent the absorption of undigested proteids may take place in the animal body and how much harm can be wrought in this way, we are not able to say. It is more than probable that the great susceptibility of the infant to bacterial products formed in milk is due to the fact that during this period of life the intestinal walls permit the passage of proteid bodies, to which the same structure in the adult is impervious. When peptons and albumoses are injected directly into the blood they act as powerful poisons. [Vaughan and Novy may not have been aware of the concept of immune reaction.] They destroy the coagulability of the blood, lower the blood pressure, and in large quantities cause speedy death. The lassitude and depression following a full meal, especially one rich in proteids, is attributed to the absorption of peptons, but so far there is no scientific evidence bearing on this point.[53]

(2) That certain secretions and excretions of the human body are poisonous when brought into contact with tissues with which normally they have no relation, is well known. We have already referred to the action of normal bile when brought in contact with the pancreas,[54] and that the bile acids have a hemolytic action when absorbed into the circulation is a fact which has long been known.

(3) It is the function of certain organs of the body to prevent the passage of certain substances into the general circulation. In other words, it is the duty of certain groups of cells to protect other communities from harmful agents. The rich therapeutical results which have followed experimental investigations of the relation of the thyroid gland to myxedema and cretinism are illustrations under this head. The probability is that myxedema is a form of mucinaemia, and that the introduction of an excess of mucus into the other tissues is prevented by the normal action of the thyroid gland. This protective action of certain glands is manifest both in certain internal and external secretions.[55]

(4) That the undue retention of excrementitious substances frequently leads to disturbances of health, is well known. The absorption of effete matter from the intestines and the retention of substances which should be eliminated by the kidneys may lead to disastrous results. We have only to mention as an illustration under this head the retention of urates in the causation of gout, and the absorption of bile in cases of obstructive jaundice. Our studies of the leucomaines have shown that small amounts of substances more or less toxic are constantly being formed in cellular metabolism, and the undue retention of these leads to disease.

(5) That certain cells in the body fail to adjust themselves to general alterations taking place in other organs at certain periods of life, is quite evident. So true is this that the physician recognizes the fact that there are certain periods, such as that of puberty or the climacteric, which are accompanied by special dangers to health and even to life. The most plausible explanation of this is on the supposition that in the special disturbances of certain organs other parts of the body fall out of harmony, and the parts no longer work together smoothly.

(6) Under conditions but little understood at present certain cells of the body fail to utilize certain food-stuffs. This is true, for instance, in certain forms of diabetes. The cells which are accustomed to absorb and utilize the sugars find themselves unable to accomplish this duty, and the unused sugar acts as a poison to other tissues. [The discovery of insulin would not be for another two decades.]

(7) Active poisons are sometimes formed by certain cells in the body. In this way we account for the presence of certain of the more highly toxic leucomains and some of the more poisonous acids.[56]

Among that challenges Vaughan directly addressed during this period was that of rabies. While isolated cases of rabies—animal and human—had appeared periodically in the state, in 1900 Vaughan dealt with one case in his own back yard, so to speak.

> *Alleged rabies in Ypsilanti city, Washtenaw County.*—August 25, 1900. J. [James] Huestin, M.D., of Ypsilanti city, Washtenaw county, wrote the following letter to this office [Michigan Department of Health]:—"You will see that I had a case of rabies to care for. Dr. Vaughan saw the case with me and after death secured the cord and will make a full report. It was the usual terrible death. Mr. Tuttle lost a horse. The same day his dog bit him. His son-in-law skinned the horse not thinking it was rabies. The horse had all the symptoms of the disease. I did not see the horse so may have another case. After this disease developed there was no time to take a Pasteur cure. We should have some place in this State for Pasteur treatment if that has proven anything. What can be done with the thousands of dogs unmuzzled, and many running with all the symptoms of rabies?"

On August 27, Secretary [Henry B.] Baker responded. "I have notified Hon. J.H. Brown, President of the State Livestock Commission. Battle Creek, Michigan, and suggested to him that the subject have immediate attention. Do you not think it would be advisable to recommend that the men who skinned the horse that died of rabies, visit a Pasteur Institute immediately."[57]

Vaughan subsequently recommended to the University of Michigan Board of Regents at the March 1903 meeting that a Pasteur institute be established at the university. The institute was to be associated with the Hygienic Department, with the intent of treating patients possibly exposed to rabies. After unanimous approval, the board appointed Dr. Thomas Cooley, assistant professor of hygiene, to be in charge of the institute. The board appropriated $2,500 to pay salaries and maintenance associated with the institute.

The results of Vaughan's request were reported to the state medical society:

> Dr. Vaughan reported as a special committee on rabies. He stated that two or three years ago rabies was very prevalent in New York, and it gradually spread partly through Ohio into Michigan. The first case here under his observation was near Ypsilanti, where a man died of the disease. From that time, rabies has spread to every part of the lower peninsula of Michigan, and is now very prevalent among cattle, hogs, and other domestic animals. Many dogs and children have been bitten, and one child died of the disease at Saginaw. At the recent meeting of the board of regents of the university, Dr. Vaughan recommended that a Pasteur institute be re-established at the university, which was done. There are already six patients being treated there, five of whom were bitten by dogs

known to be infected with rabies. It takes three weeks to treat patients. Residents of Michigan are treated free of charge, but their room and board are not supplied by the university free. The doctor [Vaughan] thinks the $2,500 which the university appropriated to maintain the institute for a year is money well invested. The loss in Michigan from cattle alone has already been several thousand dollars. The laboratory has been applied to for virus with which to treat animals as well as persons.[58]

The Pasteur institute at the university was the first of its kind established west of New York, and remained so for many years.

The Vaughan "Tank"

As Davenport has pointed out, the protein poison, whether ptomaine or leucomaine to use Vaughan's words, was the major feature in Vaughan's understanding of disease.[59] In testing these theories using extracted or purified quantities of bacterial extracts, Vaughan often found it necessary to grow large quantities of such organisms in what today would be known as batch cultures. There was of course an element of danger in the growing and handling of literally pound quantities of pathogenic and potentially deadly microorganisms. While Vaughan was cognizant of the dangers, he felt that with proper handling and techniques the process could be carried out safely.

In order to carry out this work Vaughan developed what became known as the "Vaughan tank." Design of these tanks could be described as ingenious. The copper tanks were some ten feet in length with a width of two feet and four inches deep, producing an area of approximately twenty square feet.

> A trough around the edge one inch deep has a cover which, when lowered into place, rests in the trough. This tank is supported by an iron frame of glass piping, the legs of which rest on rollers, so that the whole may be easily moved about the room. An inner tank, two inches shorter and two inches narrower, also provided with a trough that runs around the edge, sits in the large one, and is supported two inches from the bottom of the larger one by iron cross-bars. The bottom of the outer tank and the seal trough on the edges are filled with water. The seal trough of the inner tank is filled with glycerin. Both lids are raised and lowered by iron ropes passed through pulleys fixed in the ceiling. The iron frame supporting the tanks may be of any desired height. In our incubating room we have a nest of six tanks, three of which are on frames four feet high, and three on frames two feet high. This economizes space as the lower ones can be rolled under the higher ones. Both lids are supplied with vent tubes which are plugged with cotton in sterilization. Twenty liters of three percent agar is placed in the inner tank; both lids are lowered into their respective troughs, and with three large gas burners at full blast underneath, the apparatus is a sterilizer. After three sterilizations on successive days the medium is inoculated by pour-

ing a liquid culture through the vent tubes in the lid of the inner tank. Then with upper lid lowered into the water trough and gentle heat, which may be controlled by a thermoregulator, it becomes an incubator. With a number of tanks in a small room it is better to heat the room to the desired temperature, thus regulating the heat, than it is to heat each tank separately.[60]

Once the incubator system for growing bacteria in batch culture, to use the modern term, was developed—a necessity for the study of significant quantities of bacterial protein—Vaughan and Novy could now proceed to test their theory that protein poisons were actually byproducts of protein metabolism.

> In the first place, in order to study the chemistry of bacteria, it was necessary to obtain bacterial cell substance in large amount. After many trials and many failures, I succeeded at last in constructing a tank in which bacteria are grown.... I have followed the procedure of the scientific farmer and have rotated my crops. I ordinarily grow on the tanks, first, a crop of pneumococcus. I grow this at 38°C. It reaches its maximum growth in about four days. This is removed, and the same culture medium is disinfected, or again, sterilized, and typhoid planted on it. The typhoid bacillus grows as well after the pneumococcus as it grows on a fresh culture medium. I grow typhoid bacillus for fourteen days at ordinary room temperature; then I take that off and grow colon bacillus (*Bacterium/Escherichia coli*). Two or three crops of colon bacilli. Two crops usually pay. The third crop, as a rule, does not. Then, after growing colon bacillus, I grow some saprophytic organism, and in this way get from my tanks the greatest use possible. I may say that it costs $75, or did some years ago; I suppose the cost now would be two or three times that to load the tanks, and, consequently, I want to get as much as I can.
>
> I run six of these [twenty square foot] tanks at a time, making 120 square feet of germ substance. You know bacteria are generally grown in test-tubes, and you can see the difference between growing 120 square feet of germ substance and growing the small amount that we could get in a test tube. These bacteria are taken from the tanks. You simply have to loosen the growth, put on a water pump and pull it over into any receptacle you want to. Then all extractive and soluble material is removed. We extract with alcohol the first three days, with ether for four days. In this way we get all the extractives and fats out of the cell substance. The result is we obtain ... bacterial cell substance, literally by the pound.
>
> A good yield of the typhoid bacillus, for instance, from these tanks is something about a pound, actually five hundred grammes [*sic*]. If you take this and powder it, rub it up first in porcelain and then in agate mortars, you obtain the cell substance, without any admixture. There is no danger in handling any of this material, if you proceed with caution. I have found that all of my assistants who were engaged in grinding up the typhoid bacilli, even wearing masks as they all did, that every one of them was poisoned the first time they ground up typhoid bacilli. About two or three hours after beginning to grind, the individual would have a severe chill, with marked aching in the bones, and this chill would be followed by a fever, which would run up sometimes as high as 106, and then would gradually subside. This apparently gives immunity; so that in the

subsequent grinding, even when care was not used by the man (I have never permitted anyone to grind without a mask)—no unfortunate result ever occurred.[61]

Vaughan's description of the course of the illness among his laboratory workers does not appear to be that of actual typhoid fever. The method of infection, whether through ingestion or respiratory, is unclear. The incubation period for typhoid fever would be on the order of days, not hours. And for sufficient pathogen to be internalized such that symptoms develop within hours, the result would more likely be fatal. Since this exposure is occurring prior to extraction of lipids, a more likely explanation for the illnesses developed by the laboratory workers would be the exposure to endotoxin, the lipid component of the Gram-negative typhoid bacilli which were being internalized despite the care exhibited. The result, once the patient had recovered, would still be a level of immunity to the organism.

The "Michigan Method" of Water Analysis

When the Hygienic Laboratory was established at the university during the late 1880s, three criteria were of primary importance in order to gain approval by the university board of regents: (1) Research into the cause of disease was to be carried out; (2) The laboratory was to carry out food and water analysis at the request of health officers throughout the state for the benefit of statewide communities; and (3) Instruction of students. When the new laboratory was completed, Vaughan established specific rules, known as the "Michigan Standard," for defining the purity of water. Contamination by organic or inorganic residues was to be below a certain minimum level, one which turned out to be impractical for the time. Vaughan later qualified his standard by indicating the minimum levels of contaminants were goals, not necessarily the actual level for water to be considered pure.

With advances in the methods of bacteriological analysis, Vaughan modified the standard by adding, "A first class drinking water will not contain more than 50 bacteria per cubic centimeter; and a first class drinking water will contain no toxicogenic organisms."[62]

Procedures Vaughan developed and put into practice during the first years of the laboratory remained largely unchanged throughout his tenure, and became the chosen method used by numerous other public health laboratories around the nation. The water was collected aseptically into sterilized containers consisting of one-gallon, glass stoppered bottles. Sterilization was either by steam or through the use of a carbolic acid, nitric acid or sulphuric

Bacteriological laboratory (Bentley Historical Library, University of Michigan [image #002052]).

acid solution. What seems obvious today when one describes sterilization was new to much of the medical community in the late nineteenth century; Vaughan related how when once he sent sterilized bottles to a physician in Chicago for collection of water for the World's Fair Commission, that physician opened the bottles and rinsed with alcohol for the purpose of completing the sterilization. (Vaughan returned the bottles, again sterilized, with the request that further sterilization not be carried out.)[63]

The method of water analysis utilized both agar plates as well as a rich "beef tea" broth. Samples of the water were inoculated onto agar plates, which were then placed at either room temperature or at 38°C. The theory was that common bacteria found in water would grow at room temperature, but not likely at the elevated temperature. Pathogenic bacteria would likely grow at the higher temperature because that is the temperature of the body. If there was no growth at 38°C, the water was generally considered free of bacterial contamination. The obvious problem here was that of numbers. If the number of pathogenic bacteria was low, the samples which were placed in the dishes might not have been contaminated; Vaughan did not consider this to be a significant problem.

The use of a beef tea medium has some similarity to methods of present day water analysis. Tubes were inoculated with different quantities of water to be analyzed, ranging from $\frac{1}{20}$ of a cubic centimeter to one cc. After incubation, small quantities of the medium were inoculated into the abdominal cavities of either rats or guinea pigs, a method no longer practiced. If the animal died, samples of blood from the heart were inoculated onto agar plates, and the colonies compared with those which might have appeared from the water samples which had been plated. Cultures common to both were then analyzed.

Vaughan defined "filthy" water as that which contained greater than 2,000 bacteria per cubic centimeter without the presence of any toxicogenic bacteria. This did not necessarily mean the water could be a source of disease, but merely that it was not to be considered safe to drink. Vaughan became one of the first to apply the concept of the colon bacillus, now known as *Escherichia coli*, as a surrogate marker for pathogenic organisms such as the typhoid bacillus. He had recognized that in some outbreaks of typhoid, the level of bacterial content was relatively low, and the only toxicogenic organism they could isolate was the colon bacillus. Thus if the colon bacillus is absent, it is likely the typhoid bacillus is absent as well. "The trouble comes when the water is found to contain typical colon bacteria. I think that all waters containing this organism should be regarded as suspicious. Its presence indicates contamination, and if the colon germ gets into the water, the typhoid bacillus may find access to the same water."[64]

The Nature of Protein Poisons

Since determining the nature of the bacterial protein poison was among the goals of this work, Vaughan described the process of purification of the bacterial product and the experimentation which followed.

> Now you understand that this is dead bacterial substance. When you put some of this under the microscope, you will find the bacilli are very much as they appear in fresh cultures; in fact, you could not tell the difference, except that some of them are more or less broken mechanically by the grinding. I have bacilli grown over twelve years ago in a powder, and putting some of these under the microscope, they appear as fresh bacilli would.... I have found that if I took these dead bacilli and injected small amounts into animals, that they made the animals ill, and if sufficient quantity was used (and the quantity with most bacteria need not be large), it kills.

Now, there are some very interesting things about the poisonous action of this dead bacterial cell substance. I attempted to extract from these dead bacilli the poison and get it in soluble form. I worked over this for several years before I succeeded in doing so, but at last I hit upon a method which I have no doubt can be greatly improved, but it works very satisfactorily.... The poison passes into solution, while the non-poisonous part remains undissolved. Now, this poison is a pretty active poison.

Depending upon the alcohol—ethanol?—it is likely the lipid/endotoxin portion of the bacteria would be soluble in the solvent, and hence separated from the non-poisonous portion of the dead organisms. Vaughan also observed that the poisonous substance was resistant to enzymes such as pepsin and trypsin. This also would suggest the toxic portion is not protein, and quite possibly lipid. Again the evidence is that Vaughan was observing the result of endotoxin.[65]

Having found this poison in pathogenic bacteria, quite naturally, I tried non-pathogenic bacteria, and I was quite surprised to find just as much of the poison in non-pathogenic bacteria as in pathogenic bacteria. In fact, that bacterium which has furnished me the largest amount of poison is the bacillus prodigiosus, which is a non-pathogenic bacterium. [*Bacillus prodigiosus* was likely what is now known as *Serratia*, also a Gram-negative cell which would contain endotoxin.] Having found this poison, which I have called the protein poison, in both pathogenic and non-pathogenic bacteria, quite naturally I went to work to find out whether animal protein contains this poison. So I worked with the white of egg, with albumin, and I worked with blood serum, with the fibre [sic] of meat and various other animal proteins, and I found the same poison in these proteins that I found in the bacteria.

When I reached this point, I may say that I got into a stage of fright. I wondered whether my vegetarian friends were right, after all, and that we were wrong in eating animal proteins. So with somewhat feverish haste I set every resource of the laboratory at work to find out whether vegetable proteins contained this poison or not; and I was very much gratified to find that they contained just as much of the poison as animal protein does. So, so far as the poison is concerned, it does not make difference whether it is a bacterial protein or an animal protein. There is a poisonous group in all proteins; it is apparently the same group. I say apparently the same group. So far as its physiologic action is concerned, it has the same effect upon animals and upon men.... There are slight differences, chemically, no doubt, between these poisons from different sources; but all proteins contain a poisonous group. This I worked out some years ago, and I may say that it has been so widely and so universally confirmed that I think I can say with absolute certainty today that every protein contains a poisonous group. [Charles] Nicolle in France, was the first foreigner to take up this work, and he confirmed it. Later [Franz] Friedberger, in Germany, and various men in nearly every part of the civilized world have worked at it, and all have reached the same result. So there is no question now that protein contains a poisonous group.[66]

During these years of the first decade of the century Vaughan attempted to reconcile his view of the production of the protein poison with development of immunity. Much of the preliminary work by scientists had been carried out in Germany either in Koch's laboratory or in the laboratories of his former students or associates. It was routine—one might even add the word unprofessional—in this aspect of German chauvinism which denigrated the work carried out elsewhere. "It is unfortunately true that much of the scientific work done in America must go to Germany and be approved before it is accepted by other Americans [or other Germans for that matter].... The German has so long been accustomed to stamp his products as 'Made in Germany,' that much of our scientific work comes back with this stamp upon it."[67] Vaughan's approach to immunity could arguably be considered a counterpoint to German ideas, particularly with respect to his views of protein sensitivity, and in an extreme response to the same, anaphylaxis (a term Vaughan did not accept).

But how did Vaughan explain the generation of the protein poison? "In the healthy man all proteins are broken down into amino acids by the ferments of the alimentary canal, and that these amino acids are, either through absorption or directly thereafter, resynthesized so as to form the proteins which are characteristic of man's body. The precipitin test has demonstrated that every species of animal has its own specific protein bodies. Every albuminous body has its own poisonous group. Peptones injected into the blood act as poisons; therefore the peptone group contains a poisonous molecule, and it is this poisonous molecule in the peptone group which we have succeeded in partially isolating. The symptoms induced by this protein poison are marked and characteristic. They divide themselves into three distinct groups. Soon after the injection of a minimal fatal dose in one of the lower animals there is evidence of peripheral irritation.... In man it is characterized by itching and by an erythematous eruption which begins about the place of injection, and rapidly spreads over the body. In the second stage the animal lies in a lethargic condition, with rapid, difficult respiration.... The third stage manifests itself by clonic convulsions ... becoming more and more violent, until death results. After reaching the convulsive stage recovery is rare. The symptoms are produced by the injection of protein poison, whether obtained from bacterial, animal or vegetable proteins.... When the dose is excessive the first and even the second stage may not be observed. The animal is speedily thrown into a convulsion, and death results in a few minutes.

> We have studied this protein poison and its effects upon animals when the phenomenon of protein sensitization, improperly called anaphylaxis, was discovered. All will understand that protein sensitization is demonstrated by injecting

a protein, any protein, into an animal and waiting for a certain length of time, or until the animal becomes sensitized, when a second injection into the same animal causes the symptoms which I have described, in the same order as observed when the protein poison is administered, and that the final effects are the same. Comparing the phenomena of protein sensitization with protein poisoning [May] Wheeler and I in 1907 offered the following explanation of protein sensitization: When a foreign protein is injected into an animal it must be disposed of in some way. Unless introduced in large amount it is not eliminated by the kidney. It soon disappears from the circulating blood and is deposited in various tissues, the exact place of deposition depending upon the kind of protein injected and the species of animal. In order to deal with this foreign material certain body cells develop a specific proteolytic ferment, which splits up the protein injected, and no other. The first dose is gradually split up, and consequently produces no recognizable effect upon the animal. When a proper interval of time is allowed to elapse before the second injection, this new ferment, in the form of a zymogen [inactive form] is stored up in certain cells of the body, and when the second injection of the same protein is made this zymogen is activated, and converted into a ferment which splits up the injected protein with great rapidity, setting free the same poison which we obtained by splitting up proteins with sodium hydroxide in absolute alcohol. This explanation of the phenomena of protein sensitization was published by Wheeler and myself in 1907.

More recently we have attempted to use the knowledge which we have gained in the study of the protein poison in the explanation of many of the phenomena of immunity and of disease. The essential difference between egg-white and the typhoid bacillus is that the former is a non-labile, dead protein, while the bacillus is a labile, living protein. If egg-white could grow and multiply after being introduced under the skin, or into the blood of an animal, it would be just as dangerous to prick a finger with a needle moistened with this relatively harmless, bland protein as it would be to inoculate oneself with the anthrax bacillus. As early as 1907 Wheeler and I held that protein sensitization and bacterial immunity are one and the same thing. In sensitization the animal dies on the second dose. In immunity the animal survives the second dose. Sensitization and immunity are therefore apparently antipodal, but are in fact the same thing. A man drinks water containing the typhoid bacillus, and he does not develop typhoid fever that day, nor the next. He passes through a period of incubation, which in typhoid fever is somewhere about eight or ten days. During this time the typhoid bacillus is multiplying in his body in great numbers, and in doing so it is converting his proteins into typhoid proteins. Suddenly the period of incubation stops and the disease begins to manifest itself. The period of incubation stops when the body cells have become sensitized to the typhoid protein, and begin to break it up. From that time on the fight is between the living cells of the body with the ferment which they pour out, and the bacilli.[68]

The role of the putative ferment, first proposed as stated in June 1907, is the most important of the underlying mechanisms in Vaughan's hypothesis. In order to follow his argument one must make two assumptions. First, all proteins, whether animal, plant or bacterial, are composed of both a harmless portion and a poison. Cleavage of the protein liberates the poisonous portion. Second, a specific ferment is produced in certain cells following exposure to that protein, and is stored in an inactive form, a zymogen, until released from the cell. When the host is subjected to a second exposure to that identical antigen, the zymogen becomes activated while being released from the cell, and proceeds to cleave the protein and release the poison. But the foreign protein must have the ability to multiply, to build to a sufficient concentration in order to contain sufficient poison to cause disease; this argument is the basis for the difference between the egg-white analogy and anthrax bacilli expressed above. Fever, as Vaughan's argument continued, is also the result of that protein cleavage. But as with any scientific hypothesis, it was subjected to testing:

> It occurred to us that if this theory be true we might demonstrate it by repeated injections of small quantities of some protein body, and determine what effects such injections might have upon body temperature. In these experiments we have used egg-white principally because we wanted to get away from cellular structure and from the supposed influence of life. We wanted to take a dead substance. Of course in doing so we recognized the fact that egg-white does not grow and multiply in the body, and consequently we must keep up the supply by repeated injections.... Suffice it to say that by varying the size of the dose and the interval between the doses one can induce in the lower animals any kind of fever that one wishes. One can place an animal in a typhoid condition, and by repeated injections keep the animal in this condition with a temperature identical with that of typhoid fever for days and weeks. On the other hand, by more frequent injections one can induce in a rabbit an acute, fatal fever, terminating in a few hours; or, by again varying the size of the dose and the interval, one can secure at will the picture of remittent or intermittent fever. Fever, therefore, results from the introduction of a foreign protein into the body, the sensitization of the body cells to that protein, and finally the cleavage of that protein by the ferment elaborated by the sensitized body cells. Now in nature practically all the proteins that find their way into the body undigested are living proteins, in the form of bacteria or protozoa. They grow and multiply in the body, without materially disturbing for the time being, the life of the individual. This continues during the period of incubation but when the body cells have become sensitized and begin to split up the foreign protein the period of incubation ceases and that of disease begins.
>
> Death from any of the infectious diseases is due to one and the same poison, and that poison is a constituent of the protein molecule. Symptoms vary in different diseases for two reasons: In the first place, the foreign proteins have different predilection places in the body in which they are deposited. In the second place the ferment which splits up these foreign proteins is specific for different diseases.... The typhoid bacillus prefers the mesenteric and other glands; the pneumococcus is deposited generally in the lungs, though it may be found in the intestinal walls. The meningococcus finds its favorite place for growth and development in the coverings of the brain.

> From what has been said it must follow that fever on the whole is a beneficient process. It is one of the phenomena of parental digestion of proteins. The foreign protein has gotten into the body, is growing and multiplying, and in doing so is utilizing the proteins of man's body. It must be destroyed, and the body cells pour out a ferment which digests the foreign protein. This is nature's way of disposing of the foreign material, and it is apparently about the only way that nature has of doing it.
>
> That fever does result from a fermentative cleavage is shown not only by the facts which I have already enumerated, but those which we have learned in combatting fever. Nearly all, if not all, of the anti-febrile agents which have been employed in medicine are anti-ferments, and they lower the temperature by retarding the process of protein cleavage. Both natural and acquired immunity, apart from toxic immunity, may be explained by the facts as stated above. In natural immunity the foreign protein is either unable to grow and multiply, and this means that its ferments are unable to split up the proteins of the body, or the ferments of the body split up the invading protein before it has time to grow and multiply.[69]

Vaughan's view of acquired immunity followed the same logic. Exposure to the pathogenic agent, whether bacterial, protozoan or viral—the concept of viruses as infectious agents was developing during these years—produced a ferment within specific cells, and which is ultimately liberated upon future exposure. Other than the implication that a ferment represented some form of enzyme, Vaughan's understanding of the term remained vague.

> Acquired immunity is explained by the fact that the first attack of the disease, or inoculation with a modified virus, develops in the body cells a ferment which is stored up, and which on a second injection of the same protein, acts rapidly, and effectively, and splits up the invading virus. In vaccination for smallpox we use a virus modified by its passage through the cow. This modified smallpox virus develops in the body cells a ferment which is capable of splitting up the smallpox virus, and the next time this individual comes in contact with a smallpox patient, or receives the smallpox virus, it is split up and destroyed before it has time to grow and multiply. This also explains the beneficial effects that undoubtedly have been obtained by the various vaccines now so widely and often so unintelligently used.... It seems to me that our work upon the protein poison furnishes us with facts, by means of which we are able to explain many of the phenomena of immunity and disease.[70]

In understanding the logic Vaughan utilized in proposing a role for the ferment, it is helpful to follow the background of the story. The concept as applied to any protein, living or not, was the culmination of several years of work. The protein which was "split" was that of egg-white, since this would avoid any possible unique role played by a pathogenic organism. The protein was prepared and split using a hot alcohol-sodium hydroxide mixture; the poisonous portion remained soluble in the alcohol while the non-poisonous portion was precipitated out.[71] The essentials of the study were as follows:

The first injection with the poisonous portion produced no obvious ill effects in the animal (guinea pig) regardless of the dose. It did result, however, in sensitizing the animal. After an interval of some ten to twelve days, a second injection of the same material prepared from egg-white resulted in death of the animal within some twenty to forty minutes. When the non-poisonous portion of the egg-white was tested in a similar manner, sensitization did not result and the animals generally suffered no ill effects; the non-poisonous portion, however, did produce sensitization to unbroken egg-white. In Vaughan's opinion this

> demonstrates two things very clearly. First, it shows that our separation of the poisonous and non-poisonous portions is an actual and complete separation. If there remained in our non-poisonous portion any unbroken egg-white it should sensitize to itself, and the second dose, especially with the large quantities that we have used, should develop some symptoms even if it does not kill, for we have seen that one-fourteenth of the fatal dose of the protein develops to a marked degree the first and second stages [irritability of the animal and difficulty in breathing]. In the second place, it shows that only when the second dose contains the poisonous group does it develop symptoms. This, along with the fact that the symptoms induced by the split-off poison and those that follow the injection of unbroken egg-white into a sensitized animal are identical, proves conclusively, to us at least, that sensitization consists in rendering the animal capable of splitting up egg-white and that this cleavage in the animal yields the same products that we obtain by our artificial method in the retort.[72]

Vaughan was adamant about receiving proper credit for his "discovery." In the August 3, 1912, issue of *The Journal of the American Medical Association* an editorial titled "Some Features of Anaphylaxis" was published, giving credit for the idea to the German scientist Friedberger:

> The phenomena of anaphylaxis have given evidence of a power of adjustment or response of the organism to certain specific conditions little short of marvelous. The final explanation of what takes place within the tissues involved when an animal becomes sensitized by the injection of a millionth of a gram of egg-albumin still remains to be determined. It has been most baffling, to say the least, to find it possible to induce conditions of profound shock and physiologic unbalance by supposedly inert proteins in quantities far smaller than those which characterize the dosage of our most potent drugs and familiar poisons.
> Friedberger has ventured an explanation of the mechanism of these responses which has received wide-spread notice. According to this the injected protein or sensitizing antigen induces the formation of an antibody which has the properties of an enzyme. When a reinjection of the protein is undertaken after a suitable intervening period it becomes digested by the proteolytic antibody; and the digestive reaction is assumed to give rise in these cases to intermediary chemical products which are the real toxic agents in the initiation of the anaphylactic seizures.[73]

Vaughan felt the need to respond.

Under this heading [Some Features...] an editorial in the *Journal* gives Friedberger the credit of having proposed the theory that in anaphylaxis an enzyme is formed which splits up the protein on its second injection and that the harmful effects are due to the split products resulting from this cleavage. Friedberger has made many important contributions to our knowledge of anaphylaxis and some of these strongly support the theory of the formation of ferments, but he was not the first to propose this theory and I do not think he would make such a claim for himself. The ferment theory was first proposed by Vaughn and Wheeler.... The American investigators are generally given the credit for this theory in Germany, but not in your editorial.

Not only were Vaughan and Wheeler the first to propose this theory, but it was in the laboratory of hygiene of the University of Michigan that the actual existence of such a ferment in the sensitized animal was demonstrated and its cleavage action proved.[74]

Vaughan and Anaphylaxis

While Vaughan recognized early in his career in medical research the significance of what is now known as anaphylaxis or anaphylactic shock and the consequences to the animal suffering from the condition, his viewpoint as to the mechanism differed little from his theories dealing with the formation and release of ferments upon exposure to any protein.

The fact that animals which have once received an injection of protein are liable to sudden death after a second injection of the same kind has been known for many years. Ever since the opening of the Hygienic Laboratory of the University of Michigan (1888), animals once used have been segregated and kept in cages marked "used animals," which indicated that conclusions could not be safely drawn from results obtained when these animals were employed a second time. In the standardization of diphtheria toxin it soon became evident that the guinea-pig that survives one test could not be relied upon in a second one. In the late nineties, Parke, Davis & Co., large manufacturers of antitoxin, ascertained this fact and offered to supply the Hygienic Laboratory of the University of Michigan with "used" guinea-pigs at a small price. The offer was accepted, but the animals were found dear at any price, as they suddenly and unexpectedly died when treated with horse serum.

Friedemann offers the following definition: "We speak of anaphylaxis when the organism, in consequence of a previous treatment with an antigen, after a period of incubation becomes hypersensitive to the same or to a closely related substance, and when this condition can be passively transferred to fresh animals by the serum or organ extracts of the sensitized animal.... It is desirable to have a clear understanding of the meaning of the terms employed when discussing the subject. The substance which induces the anaphylactic state is generally known as the "antigen." This implies that it gives rise to the production of an antibody, and the selection of this word has been determined by an attempt to correlate the phenomena of anaphylaxis with the theory evolved by Ehrlich in explanation of the production of antitoxins by treatment with toxins. In truth the "antigen" of anaphylaxis is not a toxin, nor is the new substance generated in the body of the treated animal an antitoxin. The term "anaphylactogen" is unobjectionable, since it is applicable to any substance which induces the anaphylactic state. Sensitizer is a good word, and commits one to no theory. The same is true of the term "sensibilisinogen" used by our French confreres. The sensitizer causes the body cells of the treated animal to elaborate a specific proteolytic ferment which digests or splits up the sensitizer. Again, following the nomenclature of Ehrlich, this ferment elaborated as a consequence of the introduction of the sensitizer is generally designated as the "antibody." It would be equally rational to speak of pepsin as an antibody of beefsteak, because the former digests the latter.

The theory evolved by Ehrlich in his studies on toxin immunity is the product of a genius of the highest order. It has stimulated research, which has resulted in discoveries of the greatest importance, but the attempt to explain all physiologic and pathologic processes by this theory, and to describe them in the nomenclature of this theory is unscientific. To say that anaphylaxis is the result of protein-antiprotein reaction is to talk jargon. When foreign proteins are taken into the alimentary canal they must be digested before they are absorbed. This means that their large molecules must be split into smaller ones, and this must be continued until there are no more protein molecules left. Every protein molecule contains a poisonous group, and in normal, alimentary digestion this group is rendered non-poisonous by further cleavage before absorption takes place. When foreign proteins find their way into the blood and tissues they must be digested. This is accomplished, as it is in the alimentary canal, by proteolytic ferments but the danger from the poisonous group in the protein molecule is evidently greater in parenteral [injection] than in enteral [oral] digestion. Both enteral and parenteral digestion are physiologic processes. Every living cell has its own proteolytic ferments, otherwise it could not live. When stimulated it pours out this ferment, and it does so only when stimulated. The function of a cell ferment depends upon the kind of cell elaborating it, and to a certain extent upon the stimulating substance. The proteins are the normal stimulants to cell secretions.

When a foreign protein is introduced into the blood or tissue it stimulates certain body cells to elaborate that specific ferment which will digest that specific protein. When such a protein first comes into contact with the body cells the latter are unprepared to digest the former, but this function is gradually acquired. The protein contained in the first injection is slowly digested, and no ill effects are observable. When subsequent injections of the same protein are made, the cells, prepared by the first injection, pour out the specific ferment more promptly and the effects are determined by the rapidity with which the digestion takes place. The poisonous group in the protein molecule may be set free so rapidly and in amount sufficient to kill the animal. This in brief is an explanation of the phenomena of anaphylaxis.[75]

Fever as a Product of Action by Ferment

Vaughan also theorized that fever as well is a byproduct, albeit one which is beneficial to the animal—human or otherwise—of the function of the ferment produced in response to a foreign protein. "The subnormal temperature which may occur in the course of a fever or at its termination is due to the rapid liberation of the protein poison, which in small doses causes an elevation, and in larger doses a depression of temperature. Fever *per se* must be regarded as a beneficent phenomenon, inasmuch as it results from a process inaugurated by the body cells for the purpose of ridding the body of foreign substances. The evident sources of excessive heat production in fever are the following: (a) That arising from the unusual activity of the cells supplying the enzyme; (b) That arising from the cleavage of the foreign protein; (c) That arising from the destructive reaction between the split products, from the foreign and the proteins of the body."[76]

Vaughan's viewpoints on the origin of fever changed little by the time he ceased active research just before the United States entered World War I; in fact he found support in the experimental results observed by others in the field.

> It has been known for a long time that the parenteral introduction of proteins in the animal body may be followed by fever. [Russian bacteriologist Nikolai] Gamaleia made a most important contribution to this subject. Gamaleia found that fever follows the parenteral introduction of bacterial protein, both pathogenic and nonpathogenic, both living and dead; consequently he concluded that fever is a result not directly of bacterial growth, but of bacterial destruction in the body. Indeed, he observed that attenuated bacteria often induce a higher and more persistent fever than the virulent forms. When a rabbit is inoculated with a virulent anthrax bacillus, fever develops but persists only a few hours, and then the temperature falls below the normal and death occurs. On the other hand, when the second vaccine is used on a fresh animal, fever appears and continues for three days. When a highly virulent anthrax bacillus is employed there may be no fever and death follows within six or seven hours. Gamaleia made similar observations in other infections and came to the following conclusion: "Fever is not a result of bacterial growth, but on the contrary is consequent upon a reaction on the part of the body against the bacteria and leads to their destruction." Furthermore he found that nonpathogenic bacteria, living or dead, led to the development of fever. I think that these experiments, made more than a quarter of a century ago [ca. 1880s], furnish strong support of my theory that fever is due to the parenteral destruction of proteins.... In 1909 my students and I showed that by regulating the amount and frequency of the dosage [of albumoses or peptones] we could induce any desired form of fever, acute, fatal, intermittent, remittent or continued.[77]

Vaughan's contention that fever results from cleavage of proteins introduced into the body and the interactions of those split proteins with proteins within the body was largely unchanged.

Chapter 8

The 1910s: Dean and National Service Again

In 1908 George Dock resigned as professor of medicine in the medical school, to be succeeded by Dr. Albert Walter Hewlett. Hewlett graduated from the Johns Hopkins School of Medicine in 1900, and after private practice and an appointment at Cooper Medical College in San Francisco, accepted the offer to come to Ann Arbor. Part of the reason for Dock's departure was his frustration over the amount of clinical material available for teaching as well as concerns with the ability of members of the faculty to balance research and teaching. By the 1900s many of the facilities at the university had aged, and with insufficient funding on the part of the university administration, it was increasingly a challenge to maintain the ground-breaking work for which the university had become known. Dock had urged a union between the university's medical school with that of the Detroit Medical College, now the Wayne State School of Medicine, but with little support for this proposal among the rest of the faculty Dock felt resignation was the better course; in July of that year he accepted the offer of chair of medicine at Tulane.

Vaughan revealed part of his frustration in a letter sent to Regent Walter Sawyer. "The hospital here certainly needs improvement and revision. The work in internal medicine has fallen greatly behind. Both the late incumbent [Dock] and his assistants greatly neglected the work last year, and we wish to revive and to get a man who will not be so pessimistic but will go to work in earnest and build up the work in that department. Whatever may be the result of clinical teaching in Detroit, you and I are certainly agreed that the clinical teaching here must be greatly improved, and there is no reason it should not be. I think the faculty is now a unit, perfectly harmonious and determined to pull together for the very best interest of the university."[1]

Hewlett was not the first choice of the faculty to replace Dock—the preference was for Dr. Rufus Cole, also from Johns Hopkins. Cole was not available and with Vaughan's support Hewlett was offered the positions of professor of internal medicine and director of the clinical laboratory; the appointment was approved in October 1908.

Hewlett immediately instituted changes in teaching strategies. Under Dock, third-year students attended lectures on internal medicine, while in the clinic laboratory analysis of body fluids was emphasized. Hewlett added topics such as the etiology and physical diagnosis of disease. Greater emphasis was placed on direct interaction with patients; senior students were expected to obtain medical histories from newly admitted patients and to carry out routine examinations and perform the routine laboratory tests.[2] Reflecting in part his experience with clinical laboratories in Germany, Hewlett was a strong supporter of medical educators who emphasized both teaching and laboratory research. He was keenly aware of the advancements in technology which were taking place in the medical world, gradually instituting these within the university medical facilities. For example, the use of electrocardiography for analysis of the electrical activity of the heart had been instituted in European clinical laboratories during this time. Hewlett arranged for an electrocardiograph machine to be brought to Michigan and utilized in patient diagnosis; Hewlett then developed a course for the study of cardiac arrhythmias.[3] Vaughan was so supportive of Hewlett's efforts that since Hewlett "devoted practically all his time to his work," in the absence of private practice, that he helped ensure Hewlett was the highest paid faculty member in the medical school at $4,000.[4]

Despite the presence of outstanding faculty, the fact remained that the facilities of the medical school were increasingly inadequate, a problem of which Vaughan was well aware. Unfortunately neither Vaughan nor Hewlett or anyone else among the faculty had the power to change the situation; it was the board of regents which held the "purse strings." By the end of 1915 Hewlett was seriously considering a move elsewhere to places with more modern facilities, and in December of that year announced he was accepting a position at the medical school at Stanford.

While the salary provided for Hewlett may have been adequate for his needs, the same was not necessarily the case for other members of the faculty. This was clearly among the concerns for Vaughan in maintaining a high quality faculty at Michigan, and its effects in bringing patients to the hospital facilities. In September 1915, Vaughan wrote to University President Harry Hutchins, "It is true that greater financial reward and especially more abundant provision for research work have from time to time taken from us some of our best men. This will continue from time to time. As I now write these

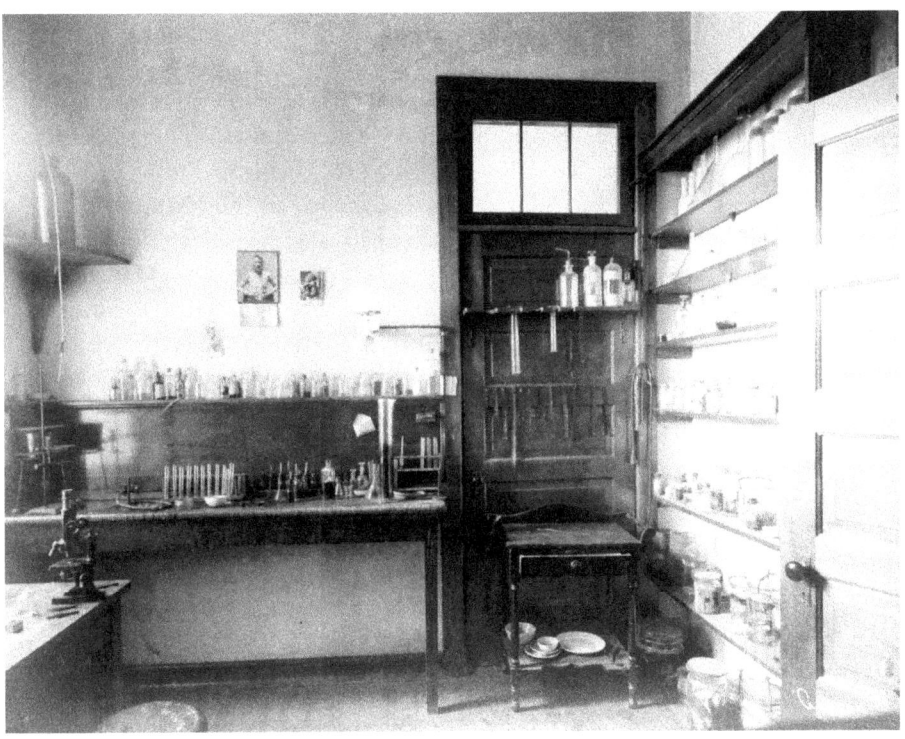

Laboratory in clinical medicine (Bentley Historical Library, University of Michigan [image #002055]).

lines, another university medical school is contemplating offering one of our best men a larger salary and possibly greater opportunity for productive research."[5]

Hewlett, Vaughan and other faculty recognized that a quality staff, including medical interns, was critical in prospective patients choosing to be admitted to the university hospital when requiring health care. In at least one instance interns were referred to as "inferior men" by the medical director, Reuben Peterson, with inadequate housing as the alleged reason. "Before then (1911) the hospital was in the humiliating position of seeing her best senior students applying for intern positions elsewhere.... While the young physician will cheerfully sacrifice much for the sake of practical hospital experience, he drew the line at the quarters and food offered him by our hospital. The best men went elsewhere for their experience. We had to be satisfied with inferior men, who were continually getting the hospital into trouble by their ignorance and lack of tact."[6] This particular problem was at least addressed, as newer housing facilities were provided by 1912.

Vaughan and the Subject of Eugenics

It is unfortunate, given our knowledge of this controversial subject today, but Vaughan was also a man of his times when it came to another subject: eugenics. Beginning in the 1890s, legislators in Michigan and other states began the introduction of bills authorizing the use of sterilization of "perverts," using contemporary language, criminals, and in late legislation, those perceived to be suffering from mental defects or deficiencies. Though few such bills were passed, Vaughan was among those in authority who campaigned in favor of certain forms of this legislation. During the years of the 1910s Vaughan presented and published some dozen lectures on the subject.

> His writings and lectures revealed a conviction that hereditary traits were a source of mental and other illnesses and cited the work and speeches of leading national figures in eugenics, such as [Charles] Davenport and [Henry] Goddard as authoritative sources. Adopting the language of good and bad "stock," Vaughan emphasized the "three-generation" theory of degenerative inheritance. Exemplary of bad stock were the "unit characters" associated with "alcoholism, feeblemindedness, epilepsy, insanity, pauperism, and criminality." He [Vaughan] explained, "All of these classes should be excluded from the list of those to whom is granted the privilege of exercising the highest, holiest, most important function of the race—parenthood.... In order to boast of good stock it is necessary to have the history of at least three successive generations. Among these there should be none of the defective unit characters mentioned above.

In 1914 Vaughan also participated in a statewide conference in Battle Creek which dealt with the subject of race betterment.[7]

This is not to accuse Vaughan of supporting other controversial precepts advocated by extremes within the eugenic movement. While he did support, and participated in, the development of tests used to measure intelligence, he was aware of how applications of such tests could be wrongly applied.

> In my opinion, another hysterical move inaugurated during the war was the undue and unjustified importance attached to the so-called "Intelligence Test" as used in the promotion of officers. I had a small part in the introduction of this procedure, but in my opinion it ran beyond bounds and secured unwarranted commendation. I thought it might be an aid in selection of those who might be entrusted with rank and increased responsibilities, but I never dreamed that it would become a dominant factor in these matters. The test, as applied in our army, was a rude measure of mental alertness, but this is only one factor in intelligence.... The so-called "Intelligence Test" as employed in our mobilization camps in the World War will need to be subjected to more crucial trials before it can justly deserve the high encomiums then and later bestowed upon it.[8]

The National Research Council

By 1916 Vaughan had additional events to deal with: war had broken out in Europe. On June 28, 1914, Archduke Franz Ferdinand of Austria, heir to the Austro-Hungarian throne, and his wife, Sophie, were assassinated by a Serbian nationalist. What had seemingly been a local dispute quickly developed into a conflict embroiling nearly all the nations of Europe as a result of various combinations of treaties and nationalism. In 1916 the United States still was attempting a modicum of neutrality, a stance becoming increasingly difficult with the loss of American lives on ships sunk by the German navy. While President Woodrow Wilson was campaigning for the upcoming election with slogans that included "he kept us out of war," he nevertheless recognized the importance of the scientific community in addressing national problems which were resulting from the European war. Among the functions of this community was to provide advice to the president in areas such as science or engineering. In hopes of making the process more efficient, a special committee was established: the National Research Council.

The umbrella organization of which he NRC was a part was the National Academy of Sciences, chartered by Congress in 1863 during the Civil War based on the idea that scientific advice could be used in the war effort. The academy came into existence when President Abraham Lincoln signed the Act of Incorporation on March 3, 1863. In addition to listing the initial membership of the academy, declaring it to be a "body corporate" under the name the National Academy of Sciences, the act outlined the mission of the academy:

> That the National Academy of Sciences shall consist of not more than fifty ordinary members, and the said corporation hereby constituted shall have power to make its own organization, including its constitution, bylaws, and rules and regulations; to fill all vacancies created by death, resignation, or otherwise; to provide for the election of foreign and domestic members, the division into classes, and all other matters needful or usual in such institution, and to report the same to Congress.
>
> That the National Academy of Sciences shall hold an annual meeting as such place in the United States as may be designated, and the Academy shall, whenever called upon by any department in the Government, investigate, examine, experiment, and report upon any subject of science or art, the actual expense of such investigations, examinations, experiments, and reports to be paid from appropriations which may be made for the purpose, but the Academy may receive no compensation whatever for any services to the government of the United States.[9]

Among the agency's first tasks was to provide methods by which the undersides of iron vessels could be protected from salt water corrosion. After considering a variety of unreliable treatments such as coating the undersides with alloys, the academy decided no immediate solution was available.

In May 1915 the British passenger ship *Lusitania* was sunk off the coast of Ireland by a German U-boat, resulting in the loss of nearly 1200 lives, including those of 128 Americans. Though Germany apologized following a strongly worded protest from the United States, it was clear to many that the U.S. would probably be drawn into the war at a future date. Two months later in July of that year, George Ellery Hale, foreign secretary of the National Academy, sent a wire to the president of the academy, William Welch, then aboard a ship traveling to the Orient, in which he strongly recommended that the academy offer its services to President Wilson. However, in addition to strongly opposing American involvement in the European war, Welch had long been a strong admirer of German culture and science, and was loath to set in motion activities which might draw America closer to war. But after Germany resumed the sinking of British ships with Americans on board, Hale pressured the academy to offer its services to the president.

Following his re-election as foreign secretary at the academy's annual meeting on April 19, 1916, Hale introduced the resolution "that the president of the academy be requested to inform the president of the United States that, in the event of a break in diplomatic relations with any other country, the academy desires to place itself at the disposal of the Government for any services within its scope."[10] The resolution was passed unanimously, and in the event that President Wilson accepted the offer, the academy was authorized to carry out any requests which might follow. A committee consisting of Edwin Conklin, professor of zoology at Princeton, Robert Woodward, Charles Walcott, secretary of the Smithsonian Institution, Hale and Welch was established to meet with the president for a formal presentation of the offer.[11]

On April 26 the committee met with the president at the White House. The discussion with President Wilson began with a general discussion of how the academy might most efficiently provide service. Wilson recommended the establishment of a committee which could organize the "scientific resources of educational and research institutions in the interest of national security and welfare."[12] In the aftermath of President Wilson's recommendation, a committee was created with the formal name of the Committee on the Organization of the Scientific Resources of the Country for National Service. Members were George Hale as chairman, Edwin Conklin, Simon Flexner, Robert Millikan, and Arthur Noyes, professor of physical chemistry at MIT. Flexner would chair the [sub]committee on preventive medicine.

At the meeting of the academy's council in New York on June 19, the organizing committee presented its recommendations.

> That there be formed a National Research Council—Hale initially referred to this as the National Research Foundation—whose purpose shall be to bring into co-operation existing governmental, educational, industrial and other research organizations with the object of encouraging the investigation of natural phenomena, the increased use of scientific research in the development of American industries, the employment of scientific methods in strengthening the national defense, and such other applications of science as will promote the national security and welfare.
>
> That the council be composed of leading American investigators and engineers, representing the Army, Navy, Smithsonian Institution and various scientific bureaus of the government; educational institutions and research endowments; and the research divisions of industrial and manufacturing establishments.[13]

The council approved the proposal presented by the organizing committee, and that date represented the formal establishment of the NRC. With additional appointments, the NRC consisted of thirty-four members by the time of its first formal meeting that September.

A letter from President Wilson to Welch dated July 24 provided official endorsement of the plan.

> I want to tell you with what gratification I have received the preliminary report of the National Research Council, which was formed at my request under the National Academy of Sciences. The outline of work there set forth and the evidences of remarkable progress towards the accomplishment of the object of the council are indeed gratifying. May I not take this occasion to say that the departments of the government are ready to cooperate in every way that may be required, and that the heads of the departments most immediately concerned are now, at my request, actively engaged in considering the best methods of cooperation.
>
> Representatives of government bureaus will be appointed as members of the Research Council as the council desires.[14]

Among the names Wilson submitted to Welch as government appointees was that of Surgeon General (Major General) William Gorgas.

The NRC was formally organized at a meeting on September 20 in the Engineering Foundation Building in New York. Nineteen members were present and George Hale was chosen as the permanent chairman. An executive committee was established, many of the members of which had been on the original organizing committee. Among the newly appointed members of the executive committee was Victor Vaughan, who would head the medicine and hygiene committee of the NRC.[15]

The NRC was actually one of two organizations established by President Wilson with the approval of Congress for the purpose of advising the president on the production and distribution of goods and services; the other was the Council of National Defense, consisting of members of the president's cabinet. While the two organizations bore no official connection with each other, the infighting both among the members of the CND and with the NRC over authority produced for a time a measure of chaotic activity; following the declaration of war in April 1917, the CNB was re-established as the War Industry Board, with a much better defined function. Meanwhile, to clarify its role as distinct from that of the NRC, a resolution was passed at the February 28, 1917, meeting of the CND "that the Council of National Defense, recognizing that the National Research Council, at the request of the president of the United States, has organized the scientific forces of the country in the interest of national defense and national welfare, requests that the National Research Council cooperate with it in matters pertaining to scientific research for national defense; and to this end the Council of National Defense suggests that the National Research Council appoint a committee of not more than three, at least one of whom should be located in Washington, for the purpose of maintaining active relations with the director of the Council of National Defense."[16] The NRC would be in charge of scientific matters for the duration of its existence.

Vaughan's duties during the remaining seven months prior to the entrance of the United States into the European conflict were confined to medical affairs. Quite often he made day trips to either New York or Washington for meetings, returning that night and attempting to continue as best he could his duties as dean. The medical issues Vaughan discussed with General Gorgas and his staff included methods of sterilizing drinking water for troops whether in camp or on the march, the problems with typhoid fever he had encountered during the Spanish-American War certainly in mind; ventilation of soldiers' barracks; clothing and rations; vaccination of soldiers against smallpox and typhoid fever and detection of disease carriers, particularly those with typhoid; an adequate supply of medicine; maintenance of diagnostic laboratories and protection of the ears against the concussions of high explosives.[17]

Medical supplies and equipment would also be needed in large quantities were Americans to go to war, and maintaining a source was another problem which had to be solved. Pneumonia was a common disease with a high mortality at the time, and the drug digitalis was often used in treating or preventing complications involving the heart. Unfortunately most of the supplies of digitalis at the time were obtained from Germany, and in the event of war, the supply within the United States would be quickly exhausted. Digitalis was produced by a member of the plant family of foxgloves, and while species of foxglove grew wild in Oregon and Washington, it was not clear whether these particular plants could prove a source for the medication. It was arranged that Boy Scouts would pick leaves from the wild plants, which were then sent to the University of Minnesota for analysis. The tests were successful, and a home grown supply of the drug became available.

Despite his aversion to war, when he returned from these meetings to meet with faculty, Vaughan apparently appeared in support of American involvement, as judged by the occasional responses from his colleagues as related in his autobiography.

On April 2, 1917, President Wilson went before a joint session of Congress to ask for a declaration of war against Germany. The measure was passed by both houses during the following four days, and on April 6 the United States was officially at war.

Chapter 9

Influenza and the Great War

With the declaration of war by the United States in April 1917 the ranks of the army increased over five-fold within months to over one million men. Among the volunteers were a number of the prominent physicians of the day, including William Welch and Victor Vaughan. Vaughan, as well as his five sons, four in the medical corps, received commissions soon after the entrance of the United States into the war. Vaughan was promoted from a lieutenant in the reserve corps to the rank of major, at the time the highest rank that a reserve officer could have, with assignment to the medical division of the Council of National Defense. "At a recent meeting of the Council of National Defense the council, at the request of Dr. Franklin Martin, chairman of the committee in charge of medical activities, authorized the appointment of a general medical board to cooperate with him in coordinating the civilian and military medical forces and advising regarding the general medical problems of national defense."[1]

Martin, a Chicago surgeon and a founder of the American College of Surgeons in 1913, and Dr. Frank Simpson from Pittsburgh, had been appointed to the Advisory Commission of the Council of National Defense the previous year for the purpose of organizing the medical profession in the event the United States had to go to war. When Martin and Simpson began their work there were some one hundred and forty thousand legally qualified physicians in the country; roughly half had the proper physical or mental qualifications. It was the job of Martin and Simpson to determine who would be fit to serve.[2]

The medical board under the chairmanship of Dr. Martin, as approved by the council, included Simpson as chief of the medical section as well as vice-chairman of the committee in charge of medical activities, and some of the most important medical practitioners in the country. In addition to Victor Vaughan, the executive committee of the board consisted of Surgeon General William Gorgas from the army, Surgeon General William Braisted from the

9. Influenza and the Great War

Medical Board, Council of National Defense: (standing, left to right): Frank Simpson, Victor Vaughan and William Welch; (seated, left to right) William Braisted, William Gorgas, Rupurt Blue and Edward Martin; Inset: W.J. Mayo, Cary Grayson and C.H. Mayo (Bentley Historical Library, University of Michigan [Victor Vaughan file]).

navy, Surgeon General Rupert Blue of the Public Health Service, Dr. William Welch from Johns Hopkins, Dr. Edward Martin, professor of surgery at the University of Pennsylvania, Dr. Charles Mayo, the incoming president of the American Medical Association, and Rear Admiral Cary Grayson.[3]

One of the more ironic announcements was issued that July by Surgeon General Gorgas on behalf of the medical board, little more than a year before the outbreak of influenza among the troops. Sanitary procedures often lacking in the war with Spain two decades earlier had been made standard by 1917 and gastrointestinal disorders were significantly reduced. Gorgas' statement reflected these changes.

> We intend to make our new military cantonments as safe as science can make them.... When a recruit has once passed his examination—which will be rigid—he may rest assured that the government will put him into as nearly an ideal sanitary environment as is found anywhere, either in military or civil life.
> We are putting the best brains in the country to work on the problem and have commissioned, as majors in the reserve, specialists in the medical and surgical branches we wish to cover. Major William H. Welch, the famous pathologist of

Johns Hopkins Hospital, is working with us at our headquarters in the Mills Building. Major Victor C. Vaughan, dean of the University of Michigan Medical Department, and nationally known as an authority on sanitation, also occupies a desk here and responds to the title of "major." We have Major William H. Mayo, of Rochester on our sanitary board, and Major Charles Mayo reports here for duty.[4]

As Franklin Martin explained,

The trained physician knows that unless certain precautions are taken dangerous epidemics, such as typhoid fever or meningitis, are almost certain to occur in the Army camps, striking often with the most surprising suddenness. If they do occur, the whole country is aroused and the entire medical force hears from it. If the proper precautions are taken, however, camp life is normal, nothing happens, and the layman does not know of, or at least never gives a thought to, the effort which was necessary to prevent the disaster. The successful Army medical officer can never point to his work and say, "Here—all these men were sick and I cured them." He can only take some dry-looking statistics and point out that in the long run few of his men were sick. It can be said with reasonable accuracy that the less heard of the medical branch of the army, the more efficient it is, because usually when much is said about it the comment is of a very unpleasant character.[5]

Among the duties of the medical board during this period was to train the medical staffs around the country, including those in nursing, with the procedures which would have to be carried out. Much of this was under Vaughan's supervision. Training included methods for the cleaning and disinfection of wounds and prevention of the dissemination of disease through respiratory secretions.

That year Vaughan was promoted to colonel in the Medical Corps—during the interim, lobbying by Gorgas and others resulted in Congress lifting the prohibition on rank—and placed by Gorgas in charge of the Division of Communicable Diseases. In his autobiography, Vaughan posited the question of when can the general health of a military organization be considered satisfactory, particularly as applied to communicable diseases? This was a modest way of asking whether the program under his supervision had carried out its job in a satisfactory manner. The challenges could not be overestimated.

Mobilization brought into close proximity men from every part of the country, most carrying some form of infectious agent. Many had never traveled far from their small towns before enlisting or being drafted, and having never been exposed to the more significant illnesses, possessed little immunity against them. Whether at the site of induction or on the troop trains carrying them to an encampment, large numbers of men mingled and exchanged not only personal information, but samples of infectious agents which they may have carried. Vaughan estimated that every troop train which arrived at Camp Wheeler in Macon, Georgia, that fall carried between one and six cases of measles. The result was that on average, between one hundred and five hundred cases of the disease erupted every day until the number of susceptible troops dropped below that necessary to sustain the disease. The admission rate in hospitals associated with the camp that fall was 534 per 1,000 men; for the army in total there were more than 48,000 admissions. Measles was the leading cause of mortality in the army during 1917.[6] It was often not just the measles which killed directly; pneumonia was an all too common sequel to the viral infection.

Pneumonia during the early years of the American entrance in the war was the major killer. Out of every one thousand men stricken with measles, forty-four developed pneumonia and fourteen died, ten times the morbidity and mortality among those not stricken with measles. Among the twenty-nine most crowded army camps between September 1917 and March 1918, the death rate from pneumonia was twelve times that in the general population. Measles is a respiratory disease, and once an outbreak has begun no measure other than quarantine, unrealistic given the conditions in camps, would stop its spread. In the movie *Gone with the Wind*, Charles Hamilton, Scarlett's first husband, dies from measles shortly after his enlistment. Young persons, who rarely encounter this disease, often chuckle during this scene in the absence of understanding how serious this infection had once been. Sanitary procedures do have an impact, however, on the fecal-oral spread of disease; the number of cases of typhoid among Americans during this same period was minimal.[7]

Despite the pronouncements from the council and the information disseminated to the public from the Committee on Public Information, conditions in the camps were far from rosy. When loved ones became ill, and in some cases died, from infectious disease the families began contacting their local representatives in Congress. Members not infrequently followed up by personally inspecting camps, especially those located in their districts. In November and December, Gorgas, Vaughan and William Welch made a tour of seven camps, including Camp Wheeler, the site of the most significant measles outbreak, and Camp Funston in Kansas, about which more will be subsequently heard. Gorgas' scathing reports about some of the conditions were not only read by members of Congress, but were picked up by the *New York Times*.

> Soldiers wearing summer clothing in Winter weather with consequent prevalence of pneumonia resulting in many deaths, lack of overcoats, overcrowding of tents in such a way as to cause disease to spread, incomplete or poor hospital facilities, the bringing into camps already full, of thousands of men who should

have been segregated under observation to ascertain if there was disease among them—these are the principal counts in indictments embraced in reports to the chief of staff from Major General W.C. Gorgas, the surgeon general of the army, concerning his personal inspection of four of the camps where soldiers of the United States are being trained for service in France...General Gorgas points out that practically all the disease is brought to the camps by incoming men, and recommends the establishment of observation camps for all newcomers so that they may be observed for such time as the division surgeon shall deem necessary.[8]

Vaughan himself later commented that he had never come so close to freezing as when they inspected Camp Funston.

Congress acted. In a series of hearings both Gorgas and Secretary of War Newton Baker were forced to testify and explain the charges of neglect towards the recruits. Not surprisingly, the Medical and War Departments attempted to place the blame on each other. While not blameless, Gorgas could refer to memos he had repeatedly sent to his superiors requesting ways to decrease the levels of over-housing. In turn Baker defended the administration's handling of the problems associated with the medical issues, even revealing that two medical officers had been court-martialed and dismissed from the army as a result of "neglect of duty," resulting in the deaths of several recruits.[9] The eventual result was that the size of the medical staff was increased and attempts were made to improve the living conditions in the camps. While the immediate problems had been addressed, an even greater challenge to the health of the men would appear in the early spring of 1918: influenza.

As epidemiologist George Soper, then a major in the Sanitary Corps, described the 1918 outbreak, "The army and navy camps suffered severely from the outset. Rarely before in the history of war has infection exhibited a more explosive character or has so large a proportion of troops been infected in camps under conditions of abundant shelter and food and freedom from the strains and anxieties of conflict. The epidemic has been attended by an unusual fatality."[10] Even the demographics of mortality differed with this outbreak. As described below, influenza had been known for hundreds of years as a potentially severe respiratory illness. But in most of these outbreaks it was the very young or very old who were most at risk. In the 1918–1919 pandemic it was the relatively young, seemingly healthy, population which produced the greatest level of mortality. The overcrowded conditions in army camps, and the large numbers of soldiers constantly in transit, certainly contributed to spread of the disease. But this would not account for either the reason why these men were the most at risk, or why it exhibited such virulence in this population.

The pandemic which appeared early in March 1918 did not at first seen unusual. Wade Hampton Frost, an epidemiologist with the National Public Health Library (now the National Institutes of Health) was among the first to recognize the unusual demographic pattern, but who also pointed out that influenza was already endemic in the population: milder outbreaks had taken place in late 1915 and early 1916. There was no way to determine at the time whether there was any relationship with the 1918 pandemic. And outbreaks of influenza were hardly unusual, as a history of the disease can demonstrate. The reason was unknown in 1918. Even the etiological agent was incorrectly described as a bacterium then; certainly physicians and scientists would have no understanding of the molecular basis for evolution of new strains.

Influenza: The Virus

Influenza outbreaks have usually occurred on a regular basis, producing a relatively mild epidemic every year or so, and a much more severe infection, often developing into a widespread pandemic, has occurred every ten to twenty years. In order to understand why these types of outbreaks take place— unlike those associated with other forms of etiological agents of disease—it is necessary to understand the structure of the virus itself. Unlike most pathogenic bacteria, viruses are strict intracellular parasites which can reproduce and spread only following the infection of a living cell. In order to enter that host cell, the target must express a receptor on its surface to which the virus can attach. Once inside the cell the viral genetic material is expressed, the virus reproduces and its progeny are released from the typically dying host cell. Since the cell receptor for most viruses is unique to either the host or to cells within the host, viruses have a limited host range for infection. Most viruses are species specific. Pets do not catch colds from their human hosts.

Influenza is an exception to some of these characteristics. Many strains of the virus can infect not only humans, but birds or swine as well. Likewise bird influenza viruses may infect other species.

The surface proteins of influenza are the reason why these viruses may infect multiple hosts. Influenza encodes two types of surface proteins, one called the hemagglutinin (H protein or H antigen), the other a neuraminidase (N protein or N antigen). The internal genetic material of this virus consists of eight segments of ribonucleic acid (RNA) each encoding one or (in two cases) two gene products: proteins, including both the H as well as the N antigens. The H antigen, named for its ability to clump or agglutinate certain types of red blood cells, is the viral surface protein which attaches to the host receptor, a carbohydrate termed sialic acid. The N antigen cleaves the sialic acid on the cell surface, allowing the virus to exit the cell. The production of neutralizing antibodies by the host immune system is necessary to block the viral infection. Since the H and N antigens are on the viral surface, these are the primary targets of the host response. When a person recovers from an infection, the presence of such antibodies will protect against a later attack.

Influenza viruses can evade the immune response for two major reasons, termed antigenic drift and antigenic shift. Antigenic drift refers to the year to year, gradual change resulting from inaccurate reproduction of the viral RNA genome. While these mutations can happen during reproduction of any of the segments of RNA, it is the changes to the surface H and N proteins which are most important, since these are the sites recognized by the host immune system. This gradual change results in the level of immunity being slightly reduced. In most cases, recovery from an infection renders the host resistant to infection by that original strain, and reduced resistance to the modified form resulting from the drift.

Antigenic shift represents a significantly greater change to the virus, in some cases creating an entirely new strain against which the population has minimal immunity. It is this form of change which results in the major pandemics which recur every ten to twenty years. There are two characteristics of the virus which allow for this form of alteration. First, the greater host range specificity of the virus means strains which infect one species may also infect a second or third species; not infrequently the same cell may be infected by two entirely different strains of the virus. Second, the segmented characteristic of the genome means that if a cell is infected simultaneously by different strains of the virus, the progeny may represent a hybrid, containing RNA segments from two different species. Often the rearrangement takes place in a third host. The creation of the deadly 1918 strain of the influenza virus may have occurred when pigs were infected with human and bird influenza strains at the same time. Or, as an alternative possibility, it may have been the rearrangement of a bird and swine virus which created the new strain. Regardless of how it came about, the strain which began circulating early in 1918 became increasingly virulent to a human population with minimal immunity against it.

Influenza Outbreaks Prior to 1918

Though definitive descriptions of what likely were influenza outbreaks date from the early fourteenth century, similar outbreaks which spread over large areas of the civilized world certainly occurred centuries earlier. Summaries and analyses of these early outbreaks can be found in the literature, albeit in the context of understanding disease among writers of the time.[11] Other medical historians, notably K.D. Patterson, describe more recent outbreaks in greater detail, or at least those which took place during the eighteenth and nineteenth centuries.[12] In addition, several authors provide detailed accounts of the 1918 epidemic within the general population of the United States, not surprising given the unique historical severity of that particular influenza outbreak.[13]

The late virologist and noted influenza expert Edwin Kilbourne suggested modern urbanization and concomitant increases in population density accounted for the alleged increase in the number and severity of influenza epidemics in the latter half of the nineteenth century.[14] Certainly there is some support for this view. But as also pointed out by Vaughan, in North America during the three hundred year period between early in the sixteenth century (ca. 1510) and the late nineteenth century, at least twenty-four widespread outbreaks of respiratory illness could conceivably be ascribed to influenza.[15] While there would have been other respiratory diseases present during that time, the contagious nature and significant levels of morbidity would suggest these outbreaks were probably due to influenza.

There is no reason to think the populations of North America, indigenous or otherwise, were unique in this regard. The urban areas of England and those on the European continent would have been even more susceptible to outbreaks of respiratory illness—to say nothing about food or waterborne disease—given the density of the populations in the urban milieu and a lack of proper sanitary conditions. Therefore what was true in North America with regard to possible outbreaks of influenza was arguably even more likely in the European populations. The time required for ships to travel the Atlantic from Europe to North America, measured in terms of weeks to months during the sixteenth through eighteenth centuries prior to development of faster modes of travel across the ocean, suggests the source of any outbreaks on either continent was likely endogenous as well; any outbreak among passengers or crews on board ships would probably have run its course by the time the ship arrived at its destination. So Kilbourne's statement in this regard, albeit referring to unnamed "authors" and not indicating his personal view on the subject, may not be entirely accurate.

Any historical account of influenza outbreaks must be speculative, the result of incomplete knowledge of the origin of disease. The very name influenza reflects this ignorance of the etiological basis of illnesses. Historically, disease was first thought to have either a theurgic origin, the role of a "god of pestilence," or later, due to the vapors in the environment.[16] As Patterson has pointed out, the rare or sporadic respiratory illness, particularly any which appeared near the beginning of an outbreak, became part of the ill-defined category of "fevers" or other varied descriptions. For a time, physicians of the period attempted to correlate changes in atmospheric conditions, or the presence of miasmas in the air, with such outbreaks. It was not unusual to include changing meteorological conditions in descriptions of these outbreaks.

The very name influenza originated from the Italian "influenz," referring to the "influence" of mysterious vapors in the air. To historians, one respiratory disease appeared pretty similar to any other, and sorting out which illnesses were the result of infection by the influenza virus from severe, self-limiting colds is difficult or impossible. Nor was anything remotely resembling the media or professional sources of present-day society in existence during those years. The first reliable medical sources did not appear until the 1700s. "Before that era, observers had to rely upon often obscure and largely uncatalogued publications, including monastery chronicles, local newspapers [the first of which, *Relation aller Furnemmen und Gedenckwurdigen Historien*, was published in Austria in 1605], general histories, and accounts of travel and exploration."[17] Some medical historians even discount any attempt at speculation of events which took place prior to the sixteenth century.

Given these challenges, and granting that any early accounts of outbreaks were subject to the beliefs and limited observations of the contemporaries of the victims, one may speculate as to which recorded outbreaks were conceivably the result of influenza epidemics. Certain characteristics of the modern disease were likely typical even then. Drawing on those characteristics roughly outlined by Beveridge, we can tentatively identify or define an influenza outbreak as one which suddenly appeared in a particular locale, spread rapidly, and then as quickly disappeared within a month or two. The outbreak likely occurred during the fall or winter months, though its appearance could extend into the warmer months as well once the disease began to spread from the source—Beveridge attributed outbreaks in 1580, 1781, 1831 to the appearance of influenza on this evidence[18]; second, the outbreak spread outwards from the source until it encompassed a much wider area, sometimes an entire continent; third, the illness was characterized by sudden onset of a high fever and respiratory involvement accompanied by severe headaches and muscle pain; fourth, morbidity rates were high, with no class or age group spared from the infection. Most victims recovered after a few days, the exception being the elderly or very young; finally, recovery from the illness did not necessarily confer immunity—however it was defined at the time—from further attacks in the future.

The characteristics of an influenza epidemic outlined by Beveridge, and indeed among most medical historians trying to catalogue early epidemics, generally assume the epidemiology of the disease would be similar to that in contemporary times. One major difference in comparing historical epidemics with those starting in the late 1800s and later would be the rate at which the disease had spread at a time when widespread travel was significantly slower and less common. We can attempt to summarize the early history of outbreaks, albeit with the caveat that anything earlier than the fifteenth century falls in the realm of speculation.

Influenza Prior to the Sixteenth Century

The sixteenth century may appear to be an arbitrary date. However, it serves as a useful demarcation for several reasons. The most obvious is that the earliest reports of what may have been outbreaks of the disease—the period prior to and concomitant with establishment of the Greek and Roman Empires—are subject to significant interpretation. Medical and sanitation practices were crude in the extreme, and there existed no shortage of infectious diseases, many of which were similar in nature to influenza. It was also during this period that accurate descriptions of what almost certainly was influenza began to appear, most notably the epidemic of 1173 in France.

The earliest description of what sounds like an influenza epidemic may be that by Hippocrates, ca. 412 BCE However a more definitive description may be that recorded by the Roman historian Livy (Titus Livius), who lived from approximately 59 BCE through the year 17 CE. Livy's written history of the Roman Empire covered from approximately the eighth century BCE through his own lifetime. Livy described one outbreak in the following manner: "A plague, however, which broke out at that time [412 BCE], and gave more alarm than it proved destructive, diverted the people's attention from the forum and political disputes to look after their families and take care of their health. It is thought the effects of the plague were less fatal than those of the sedition [uproar against the state] would have been. The city was all over oppressed with sickness, though no great mortality ensued."[19] The term "plague" in this context does not refer to the modern bubonic plague, but rather a disease characterized by rapid dissemination and undefined characteristics. Reading between the lines here, Livy seems to be describing an illness sufficient to raise widespread concern among the population but characterized by a low mortality. This would seem to eliminate waterborne illnesses such as typhoid, all too common at the time, which would have presented a significantly higher rate of mortality. While details are obviously lacking, particularly with respect to any specific symptoms, Livy's description of the outbreak is consistent with that of influenza.

Additional outbreaks prior to the turn of the eras ascribed to influenza on the basis of similar descriptions are thought to have taken place in 393 BCE coincident with the siege of Syracuse by the Carthaginians. The description is that of an undefined "plague," and again in 43 BCE in Rome.[20] An outbreak during 591–592 CE in the city of Nimes in southern France—characterized by severe headaches, sneezing and "uncontrollable yawning," according to an account by Bishop Gregory of Tours—gave rise to a superstition continued to the present day: the custom "of making the sign of the cross over the mouth when yawning, and saying 'God Preserve [bless] you' when anyone sneezes."[21] This particular outbreak was accompanied by a significantly high mortality which is characteristic of the disease. Of course the 1918 pandemic is an obvious exception, so it is not impossible that the 591 epidemic could have been influenza.

Epidemics of what appear to have been influenza have also been described for the years 826–827, 837, 876, 888–889 and 932 by various sources.[22] The appearance of such major outbreaks at ten year intervals in several cases certainly is typical of modern influenza epidemics. Still, descriptions of these outbreaks could just as simply fit any number of diseases: The epidemic of 876, also known as Italian Fever, which followed the armies of Charlemagne, produced symptoms of "pain in the eyes and cough."[23] Likewise the 888–889 epidemic in Germany was characterized by a cough and fever, symptomatic of influenza but hardly exclusive to that illness.[24] While the illness was most apparent in the human population, in some instances birds and possibly dogs were also infected, a characteristic which is certainly typical of influenza even in modern times.

The earliest accurate descriptions of what almost certainly was influenza were by August Hirsch, professor of medicine at the University of Berlin, concerning the epidemic beginning in December 1173—characterized by an "evil and unheard of cough"[25]—which encompassed Italy, Germany and England. Hirsch provided no specific basis for this conclusion in his 1881 published work summarizing the history of infectious disease. Hirsch is considered among the most important of the medical historians as a result of his compilation of numerous epidemics, not all due to influenza. He was born in Danzig in 1817 and was famous among the members of the German medical establishment for his studies of malaria, cholera and later, Plague. His three volume tome, *Handbuch der Historisch-Geographischen Pathologie* (*Handbook of Geographical and Historical Pathology*), published in a second edition during the 1880s, is considered among the most important historical works of medicine. Hirsch died in 1894.

Hirsch included in tabular form a list of some eighty epidemics which took place primarily in central and eastern Europe between 1173 and 1875. Later outbreaks in the Americas were also included; approximately fifteen were believed by Hirsch to be influenza outbreaks.[26] Hirsch later included the caveat, "Influenza always occurs as an epidemic disease, whether within a narrow circle or even confined to particular places, or in general diffusion over wide tracts of country, over a whole continent, and, indeed, not rarely over a great part of the globe as a true pandemic. It is in this last respect that influenza takes an exceptional place among the acute infective diseases; no other of them has ever shown so pronounced a pandemic character as influenza. In estimating the distribution in space to which the disease has attained in the several epidemics, it should be kept in mind that, for many of them, the records available are but defective ones, not warranting definite conclusions as to the area of epidemic distribution."[27]

Other estimates as to the number of influenza epidemics between the twelfth and nineteenth centuries varied from those listed by Hirsch. Warren Taylor Vaughan, the son of Victor Vaughan, believed at least fifty of the outbreaks could be attributed to influenza, the first in 1387.[28] In contrast, the Belgian physician Gottlieb Gluge believed the first authenticated influenza outbreak to have taken place in 1323. Gluge was born in 1812, earning his medical degree from the University of Berlin in 1835. Gluge's monograph "Die Influenza oder Grippe, nach den Quellen Historisch-Pathologisch Dargestellt," ("The Influenza or Grippe, According to the Sources of Historical-pathology illustrated"), written while still a medical student at the university, is considered the first definitive description of the disease.[29] His work was so well received that he was awarded a prize from the medical faculty. Gluge later became a Belgian citizen, even serving as the physician for the king.

Gluge was clearly describing the commonly recognized symptoms of influenza. "Sensitivity in the sagittal suture [i.e., a severe headache] and congestion in the trachea. The senses of the eye and taste are significant, if not a sign of inflammation; the eyes can be opened only with difficulty. Sight is weakened, and taste, smell and feelings mostly missing. The ear is aching. The [muscles] are suffering; it hurts the head and neck muscles, the back and shoulders feel as if stretched, bruised, broken and removed from joints. The chest is constricted, as with an iron band.... If fever is present it is mild, beginning with a slight shudder, heat and increasing in the evening. The pulse is 90 to 100, rarely reaching 120 per minute.... Breathing is fast in most patients, anxious, sometimes wheezing from the mucus of the trachea ... this inflammation increases deeper in the bronchi, with bronchitis and pneumonia.... Children up to twelve years old are affected in smaller numbers. Women will be infected ... and abortion is a frequent occurrence during the epidemic."[30]

Obviously any indication of the total number of epidemics during any particular period is a function of when the first defined outbreak had taken place; as seen above, there is a significant divergence in opinion among medical historians. The earliest use of the term *influenza*, albeit one which was misunderstood, likely dated from a description of an outbreak in Italy in 1357.[31] The phrase applied to the outbreak, *uni influenza di freddo*, "the influence of cold," was assumed to represent the actual name of the disease.[32] The term was used by Pietro Buoninsegni in his description of an outbreak which took place in Florence some thirty years later.[33] In his treatise, Gluge pointed out as well the common usage of the phrase in a more generic sense: "*Die Italienischen chroniken fuhren alle epidemien mit den worten an: 'fu una influenza da freddo'*" ("The Italian reports describe all epidemics with the words: was an influence from cold").[34]

French historian Etienne Pasquier described an outbreak in Paris in 1403 which began near the end of April; the incidence of disease in that city alone may have reached as high as 100,000 cases.[35] Symptoms included a cough and headache but there was little in the way of any further descriptions; obviously these symptoms could be ascribed to other illnesses as well, but it is the number of affected individuals which would support the premise that this is a description of influenza. The outbreak appears to have been confined largely to Paris that year; the following year the illness reappeared in regions of Flanders (Netherlands), Saxony and Thuringia.[36] It perhaps should be noted that Pasquier was trained as a lawyer, not a physician.

A similar epidemic broke out in Paris again during the years 1410 and 1411. The description is again dependent on Pasquier, the illness also referred to as *le tac* or *le horion*: "*En 1411 y ait une autre sorte de maladie dont une infinite de personnes furent touchez par laquelle l'on perdoit le boire le manger et le dormir ... toujours trembloit et avec seestoit se las et rompu que l'on ne toucherquelques parts.* (In 1411 there is another kind of disease which affected a large number of people in which we didn't eat or sleep ... trembling with shaking, tired and pain in parts which cannot be touched.)[37] Pasquier also described a violent cough as being among the symptoms, one sufficient to cause a "rupture," or an abortion in pregnant women. "Hemorrhage from the nose and bowels was frequently observed during convalescence which sometimes lasted as long as six weeks." It appears a similar disease was simultaneously occurring within the fowl populations.[38] Physicians referred to the outbreak as a *contagion generale' de l'air*, a "general contamination of the air."[39]

Another outbreak of a similar illness appeared two years later (1413) which, again as described by Pasquier, affected at least 100,000 citizens of Paris, lasting for approximately three months; the epidemic was followed by an outbreak of what was referred to as *coqueluche*, or whooping cough, the following February and March.[40] Whether these were variations of the same illness—arguably influenza rather than whooping cough—or different illnesses entirely is unclear. Symptoms resembled those of influenza. "Courts and universities had to be closed. [Victims] slept little and suffered severe headaches, pain in the kidneys and in the rest of the body. The [illness] was fatal for the elderly." (*C'etait une espece de rhume qui causa un tel enroument que le chastelet furent obligez d'interrompre leurs seances, on dormoit peu et l'on souffroit de grandes douleurs a la teste, aux reins et par tout le reste du corps; mais le mal ne fut mortal que pour les vielles gens de toute condition.*)[41] The 1414 outbreak appears to have spread at least as far as Italy, and possibly as far as the German states (Bavaria) where the outbreak was referred to as *Tonawasches fieber*.[42]

The outbreak returned again in 1427, reportedly during the Festival of St. Remigius, commemorating the Bishop of Rheims and held in October. "*L'an 1427 vers la st. remy tient un autre air corrompu qui engendre une tres-mauvaise maladie, qui l'on appeloit Ladendo (dit un autheur de ce temps-la) et n'y avoit homme ou femme qui presque ne s'en sentit durant le temps quelle dura. Elle commencoit aux reins, comme si on eust en une forte gravelle, en apres venoient les frissons et estoit-on bien huit on dix jours qu'on ne pouvoit bonnement boire, ni manger, ni dormir.... Et s'entremoequoit le peuple l'un de l'autre, disant: As-tu point eu Ladendo.*" ("In the year 1427 during the Festival of St. Remy an illness called by the authors as 'Ladendo' began with pain in the kidneys resembling 'gravel' [kidney stones], lasting eight or ten days during which we could not eat, drink or sleep. [People going to the theater] would ask, 'Have you had Ladendo?'")[43]

Epidemic of 1510

The epidemic, or more correctly the pandemic, of 1510 is widely accepted as the first outbreak which was clearly influenza. The symptoms match closely the disease at it appears in modern times. As described by Dr. Thomas Short, mid eighteenth century British physician,

The disease called coccoluche, or coccolucheo (because the sick wore a cap or covering close all over their heads) came from the island Melite [present day Malta], in Africa, into Sicily; so into Spain and Italy, from that over the Alps into Portugal, Hungary and a great part of Germany, even to the Baltic Sea; every month shifting its situation with the wind, from East to West, so into France, Britain, etc. It attacked at once, and raged all over Europe, not missing a family and scarce a person. A grievous pain of the head, heaviness, difficulty of breathing, hoarseness, loss of strength and appetite, restlessness, watchings, from a terrible tearing cough. Presently succeeded a chillness, and so a violent cough, that many were in danger of suffocation. The first days it was without spitting; but about the seventh or eighth day, much viscid phlegm was spit up. Others (though fewer) spit only water or froth. When they began to spit, cough and shortness of breath were easier. None died, except some children. In some, it went off with a looseness; in others by sweating. Bleeding and purging did hurt. [Two medical procedures in common use until the nineteenth century.] Bole armoniac [a red clay once used as an astringent] was chiefly useful, with oily lintus's, pectoral troches, and decoctions.[44]

The description also included the range of the outbreak, from Africa through much of Europe; without question the description is that of influenza. That year marked a particularly devastating period of time in Europe in other ways. Several disastrous earthquakes took place, including one which killed some 13,000 persons in Constantinople. A volcanic eruption in Iceland spewed foul soot and air over much of western Europe and Great Britain. Finally, in 1513, "the Great Mortality," now known as bubonic plague, spread through portions of Europe. There is little record of whether a comet which had appeared just prior to the events was considered a portent of evil. But it would have been understandable if the populations had perceived it as such.

Among the Italian victims of the epidemic of 1510 may have been the eight year-old Ugo Bomcompagni, the future Pope Gregory XIII. Pope Gregory of course recovered, and his decree in 1582 replaced the older Julian calendar with the Gregorian calendar in use today. While medical historians agree he was seriously ill with influenza, there is confusion as to exactly when: during the 1510 epidemic as a young boy, or during the 1580 epidemic as an elderly adult. (See below in Pandemic of 1580.)

Short's "coccolucheo" was a reference to what was not unusual for the time, the use of local terminology for naming a disease. Names for the outbreak found in other locales included "coquelicot," possibly referring to the poppies or opiates sometimes used to treat disease, or "coqueluche," with reference to the symptomatic cough; the present-day use of the term is a reference to pertussis (whooping cough).[45]

Epidemic of 1557

In 1557 another pandemic appeared, characterized by an unusually high mortality rate according to some historians, though this level of virulence is disputed by others. Whether this was actually influenza has been debated among a number of historians—for example, Beveridge ignored the outbreak—but symptoms strongly support the argument that it was indeed influenza. As in the 1510 outbreak, it may have begun in Africa, spreading to Sicily in June and into Italy by August. By that fall the disease had spread through Switzerland and France, the Netherlands and Spain.[46]

Independent descriptions of the outbreak depict at least one highly virulent disease and quite possibly more. For example, various medical historians provide different pictures of the outbreak in France that year. In Nimes (France), an anonymous physician, referring to the disease by the earlier name of coqueluche, implying whooping cough, described a violent cough and pain in the region of the kidneys. The mortality rate was relatively high according to some witnesses. Although this particular description does not include the ages of the victims, other descriptions following the dissemination of the disease throughout France indicate both children and adults were infected, though mortality was limited largely to children; whooping cough certainly could be the correct diagnosis.[47] In contrast, Johann Coyttar said in his 1578 description of the outbreak, "*Ut vix e millibus unum invenius que non eo tempore coccelucha laborarit. Quam tamen non usque adeo periculosum aut crudelem fuisse existimes vellem, ut credas multis exitio fuisse, cum certe neminem hominum (quod sciam) viderim que hoc symptomate gravatus satis concesserit.*" ("That out of the thousands it would be difficult to find anyone suffering from the 'cough.'")[48] The description is of an outbreak significantly milder than that detailed by the anonymous physician; either the perspectives differed, or we have the description of two entirely different diseases.

Thomas Short's description provides a much more detailed discussion of the symptoms as the disease presented in England or western Europe.

> In the end of September [1557], came a very cold North wind; presently after were many Catarrhs, quickly followed by a most severe cough, pain of the side, difficulty of breathing, and a fever. The pain was neither violent nor pricking, but mild. The third day they expectorated freely. The sixth, seventh, or, at the farthest, the eighth day, all who had that pain of the side died; but such as were blooded the first or second day, recovered on the fourth or fifth; but bleeding on the last two days did no service. [Bleeding, in this context, meant the method of treatment.] Slippery, thickening linctuses, were found of most service. Broths, or spoon-meats, or moist foods, were good. But where the season continued still rainy, the case was very different; for at Mantua Carpentaria, three miles from

Madrid, the epidemic began in August, and bleeding or purging was so dangerous, that in the small town 2,000 were let blood of, and all died. [From the disease or the treatment cannot be determined. Bloodletting was a common treatment for a variety of illnesses.] There it began with a roughness of the jaws, small cough, then a strong fever, with a pain of the head, back and legs; some felt as though they were corded over the breast, and had a weight at the stomach; all which continued to the third day at farthest; then the fever went off with a sweat, or bleeding at the nose. In some few, it turned to a pleurisy, or fatal peripneumony. At Alcmaria [Almeria, Spain], this year in October, raged such an epidemic, as seized whole families at once. In that small place, died in three weeks 200 persons of this mortal peripneumony. It attacked like a catarrh, with a very slow and malignant fever, bringing, as it were, a sudden suffocation along with it; then seized the breast with so great a difficulty of breathing, that the sick seemed dying. Presently it laid hold of the precordia [region over the lower chest or thorax] and stomach, and with a violent cough, which either caused abortion, or killed gravid women. Some, but very few, had continual fevers along with it; many had double tertians [recurring fevers]; others simply slight intermittents. All were worse at night than by day; such as recovered were long valetudinary [hypochondrial], had a weak stomach, and hypped [melancholy]. This disease seized most countries very suddenly when it entered, catching thousands the same moment.[49]

While the severe cough could be symptomatic of an outbreak of whooping cough (coccelucha), the inclusion of pregnant women and other adults would suggest a respiratory infection of some sort; influenza represents the most likely possibility, a highly virulent form which easily developed into pneumonia. The disease may have recurred in England through the end of the decade; Taubenberger and Morens, citing Parish registries, indicate such an excess of deaths until at least 1560.[50] In contrast, Short described an outbreak with similar symptoms on the continent only during the following year: "At Numigen, in July [1558], raged an epidemic, which spared none, and so cruel, that it carried off many the fourth, seventh, or fourteenth day. It seized with a fever, roughness or burning inflammation of the throat, and the fever continual. The sick were afflicted with a grievous pain of the head, taring cough, and constant severe pain of the loins that they could not walk; and so large a discharge by the nose as would suffer them to breathe."[51] Short does not indicate the presence of a similar outbreak in England either that year or the next.

Vaughan has implied in his history of influenza that this outbreak may have extended as far as America.[52] While this is possible—there is no reliable information to confirm such an outbreak—one has to address the question of how an infectious viral disease could be maintained aboard a ship during a voyage lasting weeks to months.

Pandemic of 1580

Mild outbreaks, at least relatively mild in the sense of few recorded deaths, may have appeared intermittently in the next decades. Though the precise nature of the outbreak is disputed, an epidemic of some sort appeared in Scotland and portions of Europe in 1562 and 1563. As described by the sixteenth century Swiss botanist Johannes Bauhin, himself a victim of the disease, "*Gravitas capitis cum dolore et defluxionibus magnis, quibus correpti sumus ferre omnis; vocant hunc morbum Galli coqueluche*" ("A great headache and discharge which affected all. The French call this whooping cough.")[53] The outbreak is not mentioned by some medical historians. The next pandemic, of which there is no dispute that it was influenza, took place in 1580. The origin of the outbreak was believed by some sources to have been in Asia, spreading west from there. Others suggested an African origin, from where it spread to the island of Malta and from there to Sicily and the Italian peninsula. Eventually it extended through most of Europe, at least as far as the Baltic Sea, England and Asia (where it may or may not have originated).[54] Among the victims of the outbreak was Anna, wife of Philip I, King of Spain. Philip himself became ill but survived. According to some medical historians, Pope Gregory XIII was also reportedly ill with the disease, though he too recovered.[55]

As said earlier, medical historians do not agree on when Pope Gregory may have been ill: during the 1510 epidemic or during 1580. Ditmar Finkler, in his discussion of influenza for Stedman's history of infectious disease, provides a convoluted listing of who did or did not support the 1580 date: "It may be mentioned in passing that Anna, the wife of Philip I of Spain [Finkler is in error, it was Philip II], died in the epidemic of the year 1580, and that the king himself, and also Pope Gregory XIII, had the disease in a dangerous manner. For this information we are indebted to de Thon, but the event was wrongly placed by Zeviani under the epidemic of 1510. This mistake of Zeviani was propagated by Ozanam and [Dr. Heinrich] Schweich, was corrected by Gluge [incorrectly referring to Pope Gregory VIII, not XIII[56]], but has again made its appearance lately in the work of Zuelzer."[57]

Stedman's history was published in 1898, so is subject to the information available at the time. Gluge's reference to Pope Gregory does not specify in exactly which year he may have become ill. "[French historian Jaques August] De Thou: *Er stimmt mit Mezeray Beschreibung fast ganz uberein, und fuhrt ausserdem noch an, dass der Papst Gregor VIII gefahrlich an ihr krank gewesen und Philip II Gemahlin an der influenza gestorben sei.*" ("He agrees with [historian Francois Eudes de] Mezeray's description almost completely, and also concludes Pope Gregory VIII was dangerously ill, and [Anna] consort to

Philip II, had died of influenza.")[58] In contrast, David Morens and his colleagues argue for the date of 1510 as when the future Pope Gregory contracted influenza.[59] Of course it is conceivable he was ill during each of the two outbreaks. But none of the sources provide any indication that this was the case.

Short again provided a vivid description of the symptoms of the disease as it appeared in England and parts of Europe.

> From the middle of August to the end of September, raged a malignant epidemic Catarrh [severe cough, so it is understandable how this could also have been interpreted as whooping cough as in the 1562 outbreak]; it began with a pain of the head, and feverish heat; some were disposed to sleep, others to watching [inability to sleep]; presently followed a dry cough, pain of the breast, haskness [huskiness] and roughness of the throat, weakness of the stomach; at last, a terrible panting for breath, like dying persons. Though the cough lasted not long, yet the panting for breath continued to the fourteenth day; some sweated, such recovered the thirtieth or fortieth day; they did not expectorate much. With some the disease went off by stool, in others by urine. Though all had it, few died in these countries, except such as were let blood of, or had unsound viscera. [It seems the physicians of the time overlooked the relationship between bloodletting and a fatal outcome of the illness.] Of the first, died in Rome at this time 2,000.... [Presumably this was when Pope Gregory became ill.] In sundry places it began with a weariness, heaviness and painful sensation; heat and horrours seized the whole body, chiefly the breast and head, with a dry cough, hoarseness, roughness of the jaws, difficulty of breathing, weakness and languor of the stomach, vomiting green bile, like juice of leeks; which symptoms increased with the disease, as the fever, cough, weight and pain of the head, pricking pain of the extremes, watching, dryness and roughness of the tongue, and shortness of breath. At the state of the disease all these were heightened, catarrh, cough, spitting. Some had swellings on the glands of the throat. In some it went off by stool; in others by urine or sweat, or bleeding at the nose. Some had spots. With some it ended with a pleurisy, peripneumony, or consumption; all recovered very slowly. The disease raged all over Europe at least, and prevailed for six weeks.... The same epidemic returned in October and November that year; then bleeding even in these places was hurtful, except when a spitting of blood, pleurisy or peripneumony attended it.[60]

Various historians have defined the outbreak using contemporary or nomenclature which reflected the symptomatic nature of the disease. For example, the vomiting of "green bile" by some victims, as noted in Short's description above, gave rise to the designation of catarrhus biliosis by some physicians.[61] Balthasar Brunner, in his description of the outbreak in the German town of Eisleben (better known as the home of Martin Luther) "Report of the Disease of the Head" (*Bericht von der Hauptkrankheit, Eisleben 1580*), emphasized that particular feature.[62] While some individual symptoms—the severe headache described by most historians for that period, the severe cough, the presence of pleurisy or characteristics of pneumonia—certainly could apply to other illnesses prevalent at the time, the spread of the illness throughout the known civilized world and the demographics of the outbreak, affecting anyone regardless of age, largely confirm that what was being described was indeed influenza.

Were there epidemics between 1580 and the first decades of the seventeenth century when the New World was undergoing exploration and settlement by Europeans? Again evidence is scarce, suggesting perhaps that while local outbreaks might have taken place, there were no widespread pandemics. One reason, arguably, is that influenza was superseded by a far more virulent disease: the "Great Mortality," or as known today, the plague. In Spain alone during the first half of the seventeenth century, the mortality associated with the plague approached one million persons. Mortality throughout the rest of Europe was no different. If influenza was overlooked, there was certainly an understandable reason. The deaths of multiple hundreds of thousands of possible susceptibles was also capable of blocking the chain of viral transmission.

Nevertheless, there were several localized and limited outbreaks of what might have been influenza: 1591 in Germany, 1593 in Holland, France and Italy, and again six years later, and in 1610–1612, "Catarrhs and disorders of the breast over all Europe."[63] The first three are included by August Hirsch in his compilations of epidemics during the years immediately after the 1580 pandemic, though only the last seems to have any form of acceptance as influenza by historians. The 1591 outbreak exhibited features unusual for that of influenza, if indeed that is what it was: *"presque tous le malades tombaient dans un delire frenetique et mouraient le huitieme ou le dixieme jour"* ("almost all the patients fell into a frenzied delirium and died the eighth or tenth day"). Sixty thousand were said to have died from the disease in Rome that one year.[64] Hirsch was not the only historian to include several outbreaks which appear to have taken place during the 1590s. Robley Dunglison, considered the "Father of American Physiology" and a personal physician to President Thomas Jefferson, compiled a similar list of historical outbreaks of illness which were likely due to influenza. In fact, medical historians such as Zeviana, Gluge and others all compiled largely identical descriptions of outbreaks. Their compilations generally corresponded to those same dates from the late sixteenth century, lending credence to their works as historical fact.[65]

Influenza Reaches the Americas

As noted earlier, Victor Vaughan suggested influenza may have first appeared in the West as early as 1557; evidence for its appearance is scarce and, as pointed out, the likelihood such an outbreak could be maintained on a long ocean voyage would also argue against that early an appearance. So when did influenza likely make its first appearance in the Americas? Arthur Hopkirk believed it may have been as early as 1627; an influenza outbreak had appeared that winter in western Europe and, according to Hopkirk, was carried to North America, the West Indies and as far south as Chile.[66]

> The first recorded contemporary description of what may have been influenza in North America was that by New England historian William Hubbard (ca. 1621–1704): In the year 1647 an epidemical sickness passed through the whole country of New England both among Indians, English, French and Dutch. It began with a cold and in many was accompanied by a light fever. Such as bled, or used cooling drinks, generally died; such as made use of cordials, and more strengthening, comfortable things, for the most part recovered.
>
> It seems to have spread through the whole coast, at least all the English plantations in America, for in the Island of Christophers [St. Kitts] and Barbadoes there died 5 or 6,000 in each of them. Whether it might be called a plague or pestilential fever, physicians must determine. It was accompanied in those islands with a great drought, which burnt up all their potatoes and other fruits, which brought the provisions of New England into great request with them, who before that time had looked upon New England as one of the poorest, most despicable, parts of America.[67]

A vivid description of that same epidemic as it developed on the aforementioned island was provided by Richard Vines, a friend of Massachusetts Bay colony governor John Winthrop, in a response dated April 20, 1648, to a (now lost) letter sent by the governor which had described the New England outbreak. "The sickness was an absolute plague, very infectious and destroying, insomuch that in our parish there were buried twenty in a week, and many weeks together fifteen or sixteen. It first seized on the ablest men, both for account and ability of body. Many who had begun and almost finished great sugar-works, who dandled themselves in their hopes, were suddenly laid in the dust, and their estates left unto strangers. Our New England men here had their share, and so had all nations, especially Dutchmen, of whom died a great company, even the wisest of them. The contagion is well-nigh over; the Lord make us truly thankful for it, and ever mindful of his mercy."[68]

Noah Webster, in a work published in 1799, likewise referred to the 1647 outbreak among the forty-four historical epidemics to which he made reference: "It will be remarked that the year 1647 when the influenza invaded America, was a sickly year in Europe." Webster continued with references to repeated outbreaks of influenza during the remainder of that seventeenth century decade: "In 1655 when the plague was epidemic in Europe, the influenza again prevailed in America. In 1658 when the influenza invaded Europe, great sickness and mortality occurred in America."[69] Webster was of course better known for his work on language and lexicology. However, the yellow fever and influenza outbreaks during the late 1780s and early 1790s resulted in his investigation of earlier outbreaks of disease in North America.[70] Webster's diary entry for April 10, 1798, noted, "Begin to write my history of epidemic diseases, from materials which I have been three months collecting." His stated goal was to "trace back the history of such diseases as far as the history of such records extend."[71]

Pandemics of the Eighteenth and Nineteenth Centuries

K. David Patterson has estimated that thirteen significant outbreaks of influenza took place during the eighteenth century, and another twelve major outbreaks occurred during the following century. At least nine of these twenty-five were believed by Patterson to be the equivalent of modern worldwide pandemics.[72] The first of these outbreaks appears to have originated in Rome about the time of Christmas 1708. From there it spread north into western Europe, particularly France and the German states by the summer of 1709. Patterson noted that the epidemic reached as far west as Ireland, though he did not appear to believe the epidemic spread to Great Britain; if it did, the outbreak was not significant enough to warrant any description.[73] In contrast, Thompson and Kilbourne believed the illness did indeed invade Britain during this period. Quoting from the 1749 description by Thomas Short, Thompson wrote,

> The year 1710 was very temperate in the general, only in the end of March were three insufferably hot days. From April the 7th to the 11th, north wind, sleet and cold; then six days excessive heat, with east wind cooled by after rains. In June several unseasonable sharp and cold winds, from which vicissitudes of weather Catarrhs and Arthritics were not unfashionable.... [Note the inclusion of meteorological conditions in describing the outbreak of illness.] March the 1st began and reigned two months, an epidemic which missed few, and raged fatally like a plague in France and the Low Countries, and was brought by disbanded soldiers into England, viz., a Catarrhous Fever, called the Dunkirk Rant, or Dunkirk Ague; it lasted eight, ten or twelve days. Its symptoms were a severe, short, dry cough, quick pulse, great pain of the head, and over the whole body, moderate thirst; sweating and diuretics were the cure. Bleeding very pernicious or fatal. This was a very moist, southerly, and unsettled constitution in England.

The author goes on to note, "Influenza prevailed in Dublin in the previous year, after a sudden transition of atmospheric temperature from heat to cold."[74] The level of mortality was not included. Nevertheless, from the description of the outbreak as well as the presence of the disease in Ireland the previous year, one may conclude the epidemic was quite likely that of influenza.

It appears a minor outbreak took place in portions of the German states during the summer of 1712. If so, it was confined to those regions, since no reports appeared in the surrounding countries. The next significant pandemic appeared in 1729 and 1730, reappearing two years later (1732). The initial outbreaks appear to have originated in Russia in the regions of both Moscow and sites on the Caspian Sea sometime during April 1729. The time lag between its appearance in April in Russia and September in Sweden has suggested to some two separate strains of influenza; more likely it was the scarcity of medical reports which created the perception of separate outbreaks.

From there influenza spread westward that fall into Poland, Germany and France, appearing in portions of Great Britain and Ireland by December. It was said that so many monks were stricken with the "flu" that religious services had to be canceled.[75] By the following March (1730) even Iceland was affected, the first time the disease was known to have invaded that island. North America was stricken during the fall of 1832, spreading along the coast and subsequently appearing as far south as the islands of the Caribbean and as far north as Newfoundland. The level of mortality was difficult to estimate but at least in some locales appears to have been substantial. During the outbreak in London in 1729 it was reported that upwards of 1,000 persons were dying each week. In 1732 the description in Plymouth, England, was "that some were seized suddenly, they fell down in multitudes, scarce anyone escaped it." Mortality in London was three times the usual level.[76]

Between the 1730s and 1780 a series of relatively minor outbreaks took place in more or less ten-year intervals. Most were limited to portions of Europe and England and do not appear to have had a particularly high mortality rate. The pandemic which appeared in 1781, however, would rank among the most widespread in recorded history, arguably surpassed only by the 1918 pandemic.

It may have originated in China or Southeast Asia during the early fall 1780; the earliest cases were reported from these regions. The disease was so widespread that several committees were established in London to study the outbreak. By the time it had run its course in 1782, tens of millions of persons had been infected; it was said two-thirds of the population of Rome became ill, and three-quarters of the population of Munich.[77] The committee established by the Royal College of Physicians reported that cases appeared on the southeastern coast of India during the fall 1781, from there spreading into portions of Russia by December. It was equally likely that the outbreak had also been brought to portions of the Urals from China. Regardless of from which direction, and quite possibly from each, the growing pandemic reached St. Petersburg by January 1782.[78]

From Russia the pandemic spread westward into Europe in April and May, reaching southern England, including London that month, and Ireland in June. By some estimates nearly 80 percent of the population of Great Britain became ill. The level of mortality was difficult to ascertain. If one utilizes the Bills of Mortality for London, the epidemiological study of numbers and causes of deaths, one observes a significant increase in persons dying from "fever." But given the high morbidity rates, even though hundreds of thousands probably died throughout Russia and western Europe, the mortality rate was probably relatively low and confined to the older demographic. Influenza returned later that decade, but numbers appear relatively low, and the cause may have been a strain similar to that which appeared in 1781.

The influenza pandemics of the eighteenth century, unlike many of those which appeared earlier in Europe, appeared to have originated in Russia, spreading from east to west across Europe. This may not have been entirely a coincidence. Patterson ascribes the movement of the outbreaks as being the result in part of political and economic changes which were taking place during these years in Russia. Under the leadership of Peter the Great from 1682 until his death in 1725, Russia expanded its economic interactions with the west, improving trade relationships and increasing physical interactions with portions of Europe; it was not surprising that influenza would follow the same routes as that of commerce.[79]

It was not only the origin and spread of the disease which underwent a change during the eighteenth century. Prior to this period, this illness was known by a variety of names: catarrhal or catarrhous fever or simply the "fever," the ague and others. The first well documented example of the use of the term influenza was that by John Huxham in 1757, "the catarrhal Fever, which spread through all Europe under the name Influenza in the Spring, 1743." As indicated by Delacy, the term had become commonplace by the time of the 1782 pandemic.[80]

By the early years of the following century medical practitioners began as well to question in depth the very nature of influenza outbreaks, particularly their cause and mechanisms of spread. This was still a period in which medical knowledge often focused on "miasmas," the influence of air or atmospheric conditions as the agent of disease. While there were those who argued in favor of person-to-person transmission, the "contagionists," the seemingly random appearance of diseases, not only influenza, seemed to contradict this belief. The microscope was rarely applied by physicians in the study of microorganisms, and medical research was more a hobby than a profession. Yet there were those whose ideas related to disease etiology were tantalizingly close for the time to what we know to be correct. "The surgeon Richard Dunning feared the disease [influenza] was not contagious because of the swiftness of its spread and its apparently very short incubation period; if it were contagious, however, he thought that his friend Edward Jenner's recent discoveries [concerning vaccination] offered hope for controlling it. Dunning thought that the cause might be an atmosphere impregnated by animalcules."[81]

As in earlier centuries, influenza outbreaks during the nineteenth century seemed to take place at regular intervals of ten years or so. At least three could be considered pandemics while interspersed with more localized, "milder" epidemics. Disruption by war certainly contributed to the first outbreak taking place during the first decade. Originating in Russia late in 1799 or early 1800, influenza followed the French soldiers as they returned from the wars in the east. The first cases in France appeared in September 1802 in Paris, and by December had spread throughout much of western Europe and Italy. Fortunately the illness was relatively mild and most deaths were among the elderly. The first known pandemic of the century took place during the early 1830s. Influenza broke out in portions of China early in 1830, spreading south to the Philippines and Dutch East Indies. In November that year influenza also appeared in portions of Russia. Whether this was the same strain which had spread through a southern route as well, or an entirely different strain of the virus, cannot be determined. But from St. Petersburg the disease spread westward into Europe. By June 1831 it had arrived in portions of Germany, France and England. As a result of commerce and travel between Britain and the United States, the illness reached Philadelphia from ships during the fall of 1831; it spread throughout the colonies during the following months. Mortality was again confined to the elderly population.

Influenza reappeared a year later, this time in a more virulent form. As was the case in 1830, the illness first appeared in Russia, traveling west into Europe. That this represented a new strain of the virus is supported both by the fact that persons who had been infected two years earlier were still at risk in the 1833 outbreak; morbidity in many cities was reported to approach 80 percent. Mortality rates were also significantly higher in many areas; deaths in England during the month of February were reportedly double the number during other years.[82]

The epidemic which appeared in 1836 and 1837 differed in two ways from that which was widespread several years earlier. First, it was milder, perhaps the result of an antigenic drift affecting the strain which was present in 1833. In addition, rather than the east to west spread which had become almost commonplace, it appeared first in the north, some have suggested in the Baltic region, Sweden or Denmark, and from there spread south through Europe. Morbidity was again high, with some cities reporting as much as 50 percent of the population becoming ill.[83]

The epidemic of 1847–1848 affected much of Europe and the Mediterranean region, and as with most of the outbreaks during the previous century, the 1836 epidemic being an exception, appeared first in Russia and spread south and west from there. Though Beveridge has suggested the disease spread to North America, there is little evidence of this being the case. After infecting two-thirds of the population of St. Petersburg in March 1847, it spread south into Turkey and from there into Egypt and the Mediterranean. The port cities of southern France and Italy encountered the illness by that fall, and by December influenza appeared in Germany and Great Britain. The outbreak in London was severe enough—one Londoner in four was reportedly ill—that it became known as the great influenza of 1847. "The great influenza of 1847 began in London about the 16th or 18th of November, was at its height from the 22nd to the 30th, and had 'ceased to be very prevalent' by the 6th or 8th of December.... According to the Superintendent of Statistics, it caused an excess of 5,000 deaths during he six weeks that it lasted.... During the worst three weeks it raised the deaths in the age of childhood 83 percent, in the age of manhood (25–45) 104 percent, in old age 247 per cent, whereas the deaths between fifteen years and twenty-five were but little raised by it, and those between ten and fifteen hardly at all."[84]

Pandemic of 1888–1889

The outbreak of 1847 would be the last such epidemic for a generation. While the disease appeared sporadically in the Americas and portions of Europe during these years, nothing approached the levels observed in earlier decades. Significant changes had taken place during the intervening years, not only in the growth of populations and methods of transportation, but in the understanding of disease by medical science in general. Populations had grown significantly, particularly in the urban areas of Europe. But it was in the area of transportation that the greatest change had taken place. Railroads as a means of transportation had their beginnings in Europe during the first decades of the century, and by the 1880s an extensive network linked most of Europe from western Russia to the western coast of the continent. Steamships carried passengers relatively quickly from port to port. Consequently, any contagious disease, particularly one which passed through respiratory secretions, had the potential to spread quickly throughout the continent and beyond.

The concept of miasmas as the cause of disease had fallen by the wayside, so to speak, by this time. A significant number of the medical discoveries had taken place in Germany, most notably in the laboratories of Robert Koch and his associates. Koch himself had isolated the etiological agents of cholera and tuberculosis during the decade, and a colleague, Richard Pfeiffer, had isolated what was thought to be the agent of influenza: *Bacillus influenzae*, or as it was more popularly known at the time, Pfeiffer's bacillus. Despite the fact vaccines had been developed against several viral diseases by then—smallpox and rabies—the idea of submicroscopic agents such as viruses as agents of human disease would only come in the future. Pfeiffer had isolated his bacillus from the lungs of influenza victims. In the context of the time—bacteria had been proven to be the etiological agents of numerous human diseases—it made sense to report that Pfeiffer's bacillus was the agent of influenza. Most physicians believed this to be true. Victor Vaughan, when attempting to deal with the later outbreak in 1918, was not among them.

Scattered cases of influenza had been reported in portions of Russia, including St. Petersburg, through much of the decade of the 1880s. But most medical historians believe the 1889 pandemic originated Bukhara, a central Asian city with a population between 80,000 and 100,000. According to the German physician Johann Heyfelder, then residing in Russia and who described the epidemic, symptoms included a high fever, nausea and vomiting. Between 5,000 and 7,000 persons were reported to have died during the outbreak in that city. But Heyfelder's description did not include respiratory complications, leading some to believe that this was not influenza. Others, including the British physician Frank Clemow, who was in St. Petersburg during the early months of the outbreak, suspected its origin to have been in western Siberia.[85] The disease followed the rail lines through both southern and western Russia and eastward towards China. The growing epidemic reached Moscow and St. Petersburg in October 1889 and Ukraine the following month.

Once again following the path of railway passengers, influenza appeared in the major cities of Europe—Warsaw, Berlin, Vienna and Paris—that November as well. The remainder of Europe, including Italy, reported their first outbreaks in December. Passenger ships carried the disease from the ports of Russia to Britain that month. At the same time, passenger ships landing in the American ports of Boston and New York, as well as in the Canadian city of Montreal, carried passengers infected with the virus. By mid-January 1890, influenza appeared in the Midwest.

It was not only the European and American regions to which the outbreak was spreading. During January through March passengers carried influenza to the major ports of Africa, while the outbreak in eastern Russia moved south through British Asia into Australia and New Zealand.

As with previous outbreaks, it is difficult to determine accurately the morbidity levels in many of these regions. But if statistics compiled in the larger cities are reasonably accurate, it is possible to produce an approximation. In the cities of France and Germany, morbidity estimates range from 33 to 50 percent of the population. In Italy upwards of two-thirds of the population became infected. The demographics were also different from those in previous outbreaks. Instead of the elderly being most at risk, it was the age group between 20 and 30 which suffered the highest proportion of both morbidity and mortality. Approximately 60,000 persons were estimated to have died in France, 66,000 in Germany and 4,500 in the Netherlands. In London an estimated 2,800 persons reportedly died.[86] Total mortality associated with the 1889–1890 pandemic was at a minimum one million persons.

Periodic outbreaks of influenza continued through the remainder of the decade. Mortality rates were not as high, and the demographics returned to that previously observed—the elderly were most at risk. It also became possible to determine the particular strains of the virus which were circulating during those years. The science of sero-archaeology, developed later in the twentieth century, utilized preserved sera obtained from persons exposed during the 1889–1890 and 1899 outbreaks. As a result of infection, these persons developed antibodies against the particular strain with which they were infected. More specifically, the antibodies would be directed against the surface proteins on the virus, the hemagglutinin (H antigen) or neuraminidase (N antigen). The data suggest that the 1889–1890 pandemic was associated with the H2N2 strain of the virus, similar to that which reappeared in 1957. The later outbreak that decade appears to have been the result of infection with the H3N2 strain of the virus, representing a newer version of the agent and one which reappeared in 1968.[87]

The Path from Pfeiffer's Bacillus to Influenza Virus

The "identification" of a bacterium in 1892 as the etiological agent of influenza must be understood in the context of the time. The germ theory of disease, the concept of which was largely developed in German laboratories, was less than two decades old. Since etiological agents of numerous diseases, including anthrax, typhoid fever, cholera and tuberculosis among others, had been isolated, grown in laboratory cultures and identified as specific species of bacteria, the assumption was that most if not all infectious disease was associated with a bacterial infection. While at least one disease, rabies, had been shown to be caused by a filterable agent, this was still a long way from addressing the existence and role of viruses as infectious agents in animals.

In 1892, Richard Pfeiffer reported the isolation of a rod-shaped bacterium from the bronchi of patients suffering from influenza. "In all cases of influenza we found a similar bacillus.... Rods were found only with influenza while investigations showed no presence in ordinary coughs, pneumonia or tuberculosis.... Over the course of the disease, only with the drying of bronchial secretions did the rods disappear."[88] Pfeiffer's analysis and interpretation were reasonable given the state of knowledge: bacilli were present in cases of the disease and apparently absent in patients with other forms of pulmonary disease. The organism was named *Bacterium influenzae*, more commonly known as Pfeiffer's bacillus.

The first clue that the organism may not have been the etiological agent of influenza came with the inconsistent isolation from patients with the illness, though even here results were often contradictory. For example, in 1915 and 1916 C.G.A. Roos at Mulford Biological Laboratories south of Philadelphia reportedly isolated Pfeiffer's bacillus from 50 to 90 percent of influenza cases. During the 1918 epidemic, he reported the isolation of the bacilli from 82 percent of the thirty-three patients examined.[89] However in a presentation before the Pathological Society of Philadelphia, Roos indicated a different result.

> The *Bacillus influenzae* can be found in every case of true clinical influenza. To isolate this organism, which is most abundant in the early stages of the disease, it is necessary to exercise care in obtaining a suitable specimen, and since growth requirements of this organism are quite high, special selective culture media, such as suggested by [Oswald] Avery [i.e., caramelized blood agar known as chocolate agar for its appearance], also Fleming, carefully prepared and adjusted to reaction are essential for successful work.
>
> The various strains of the *Bacillus influenzae* apparently do not differ in kind but merely in degree. This is indicated by the cross agglutination, absorption, and protection tests with strains isolated at different localities during the recent pandemic as well as with those from the epidemic of a few years ago—1915 to 1917.[90]

Roos interpreted the strict growth requirements necessary to maintain the isolates in the laboratory as among the reasons some physicians or researchers were unable to isolate the organisms in all cases. Still, even if the organism was present, the question remained as to whether this was the primary cause of the disease, or was the result of a secondary bacterial infection in those patients. One method to address the question was to determine whether identical strains of the bacterium were isolated, or if different isolates were actually different strains. The significance was based upon the idea prevalent at the time that identical illnesses would be caused by identical strains of the agent. Results found that a multiplicity of strains were present: nine different strains were isolated from nine different cases of the disease. "The existence of a multiplicity of races is advanced that *B. influenzae* is not the primary etiological agent in epidemic influenza."[91] The fact that the same bacillus could be isolated from the upper respiratory tract of healthy individuals who not only never had influenza, and, to their knowledge, had never even been exposed to the illness, also lent credence to the argument that Pfeiffer's bacillus was not the agent of influenza.

The etiological agent was finally isolated and demonstrated to be a virus during the early 1930s. The first clue in this regard appeared early during the 1918 human outbreak when, at the Cedar Rapids (Iowa) Swine Show in October of that year, pigs developed a respiratory infection which resembled that of human influenza. Dr. John Koen, an inspector for the Division of Hog Cholera Control of the United States Bureau of Animal Industry, compared the respiratory infections in hogs and humans and concluded they were variations of the same illness; in fact some farm families became ill with the infection at the same time as did their animals.[92] Among those who were aware of Koen's theory was Richard Shope. Shope had been raised on an Iowa farm. After earning a medical degree from the university there he joined Paul Lewis at the Rockefeller Institute in New York, where he became involved in the study of hog cholera.

In 1929 an outbreak of swine influenza took place, and Lewis and Shope decided to study that particular disease; they shortly isolated a strain of *Haemophilus influenzae suis*, a bacterium similar to that found in human influenza and formerly known as Pfeiffer's bacillus. As had been the situation with the human isolate, the assumption was this was the etiological agent of the disease. However, when they used isolates of the bacterium to infect hogs, they were unable to produce the disease.

Following the death of Lewis from yellow fever, Shope continued to repeat the experiments, this time using filtered isolates from nasal secretions, and was able to produce the same respiratory infection, albeit in a milder form. However, when Shope combined the *Haemophilus* isolate with the cell-free filtrate, he was able to produce the identical respiratory illness in the animals; his (correct) interpretation was that the two organisms, a virus and a bacterium, worked in conjunction with each other. Further evidence that the virus was the primary etiological agent was that swine which had recovered from the illness demonstrated immunity upon further exposure.[93]

Since the etiological agent of the disease in swine which resembled human influenza was shown to be a virus, it was a logical step to determine whether the human disease was likewise the result of a viral infection. One of the challenges in attempting to fulfill Koch's Postulates, specifically the reproduction of the disease in a test animal, was that up until this time no suitable test animal besides human beings could be infected with the agent. Since influenza at best is often a debilitating disease, and at worst life-threatening, human testing was generally avoided.

The problem was solved by British investigators studying dog distemper in ferrets. Since the initial symptoms of distemper and influenza in humans were similar, British scientists Wilson Smith, Sir Christopher Andrewes and Patrick Laidlaw exposed two ferrets to infectious filtrates prepared from human throat washings; both animals became ill by the third day post-infection with symptoms that resembled human influenza: fever, watery eyes, muscle soreness and a significant watery discharge from the nasal passages. Mucous membranes in the nasal passages showed significant inflammation. Additional experiments confirmed that the three scientists had indeed isolated the etiological agent of human influenza:

> A disease of ferrets [was] produced by the intranasal instillation of filtrates of throat-washings obtained from influenza patients; The disease is transmissible serially in ferrets either by contact or by the intranasal installation of virus-containing material; The infective agent has, so far, only been recovered from the nasal passages of sick ferrets; The disease was produced by five of the eight throat washings obtained from influenza patients in the early stages of the disease; Throat washings from healthy persons and influenza convalescents caused no illness in ferrets; The nasal secretions from a subject with a severe common cold caused no illness in ferrets; Human sera, particularly those from influenza convalescents, were found to contain antibodies capable of neutralizing the virus of the ferret disease; Swine influenza virus caused a disease in ferrets which was indistinguishable from that produced by virus of human origin, and the pig and human viruses have close antigenic relationships.[94]

Using samples of the swine virus provided by Richard Shope, the authors demonstrated that human sera was capable of neutralizing ferret virus. They also attempted to explain the relationship between the isolation of *Haemophilus influenzae* from patients infected with influenza. "It is probable that in certain cases this infection [influenza] facilitates the invasion of the body by visible bacteria giving rise to various complications."[95]

Influenza Pandemic in the Army: 1918–1919

In Europe, 1918 began as a critical year in the war, but particularly so for the Allies. Three years of war had largely butchered a generation of French and British youth. Russia had undergone a revolution and signed a peace treaty with Germany, effectively ending the war in the east and allowing Germany to concentrate its forces in France. It had been nine months since the United States had declared war, but American forces in Europe were still relatively few in number and their impact was yet to be felt. In March 1918 the Germans launched a new offensive, hoping to finally break the Allied lines before the Americans could arrive in sufficient numbers to break the impasse. That spring and early summer the second enemy arrived, affecting both armies: influenza.

Influenza had appeared in December 1917 at Camp Kearny, north of San Diego in California, with 537 reported cases, spreading to additional camps the following months.[96] The disease was relatively mild, much as influenza had been during outbreaks in the preceding years; mortality was low. The

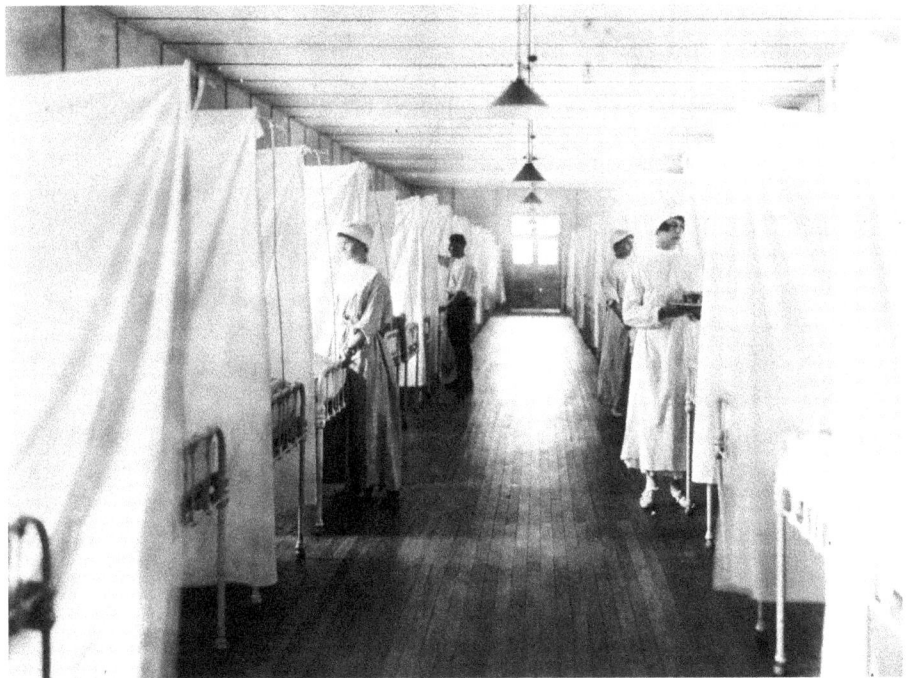

Influenza ward, Army camp hospital (Images for History of Medicine/National Library of Medicine).

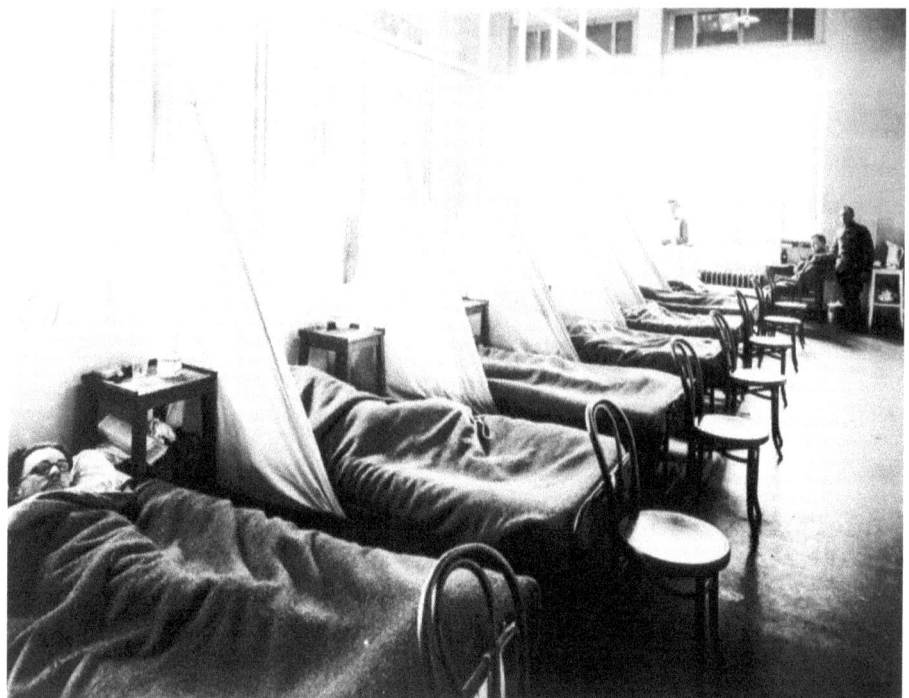

Influenza ward, Army camp hospital (Images for History of Medicine/National Library of Medicine).

500-plus cases represented some of the 32,248 reported by the army in 1917, a number low enough that it attracted little attention.

Whether troops at Camp Funston (now Fort Riley) in Kansas brought the illness with them, or the etiological agent came from the civilian population, is unknown. But during March and April the soldiers encountered three waves of the illness; the first wave beginning during the first weeks of March. The third wave occurred during late April and affected primarily new arrivals, suggesting the same strain of the virus may have been associated with each. Camp Funston that March housed some 26,000 soldiers being prepared to enter the war; eventually the number of troops at Funston would be over 50,000, the second largest cantonment in the United States. On March 4 a mess cook named Private Albert Gitchell awoke with what are generally called "flu-like" symptoms: fever, sore throat, chills and aching muscles. He reported to the infirmary, where he was immediately placed into quarantine. But by then it was too late. As a cook, Gitchell had been serving meals to the men for weeks and likely infected many of them at the same time. By the afternoon of March 11, 107 men had reported to the infirmary with the same

Influenza warning (Images for History of Medicine/National Library of Medicine).

symptoms; by the end of the week 522 men were reported ill, and by the end of March a total of 1127 men were ill. The death toll reached 46. Gitchell may have been one of the fortunate survivors, as he reportedly lived until 1968.[97] Gitchell has been occasionally tagged unfairly with the dubious distinction of being patient zero in the epidemic. With troops arriving daily, the agent could have been brought in by any number of newly enlisted men. Reports existed as well that a putrid black ash enveloped the camp as tons of manure were produced by the camp's animals. Pig and poultry farms in the (Haskell) county were in the vicinity of the camp, and since we now are aware that either pigs or birds may be the source of strains of influenza, one can as easily lay the blame, such as it is, on them.

In fact, one can make a strong case that it was not Gitchell, but local draftees who infected the camp. A physician in Haskell County, Dr. Loring Minor, had reported treating numerous patients during January and February who were stricken with what clearly was influenza. The demographics resembled those which later would be exhibited in the pandemic: mainly the young and healthy were stricken. Many of the men in the county were subsequently drafted and arrived at Funston between February 26 and March 2; the local papers even included the names of local men now numbered among the troops at Funston. Any one of them might have been the source of Gitchell's illness.[98]

At the same time as the illness broke out in Kansas, it also appeared in Europe, affecting both sides. In France it became known as the Spanish flu, likely because Spain was not involved in the conflict and was not censoring the outbreak; the only time a brief report came about was when King Alphonse XIII became ill with influenza. In Spain they often referred to it simply as the "grippe." Authorities in Spain blamed the French, suggesting the disease had been carried by the wind from the battlefields. The French were forced to move 1,500 to 2,000 men per day in May from the front lines; the British reported over 36,000 cases in May and June.[99] In Germany they called it Flanders fever. As General Erich Ludendorff, one of the commanders of the German army, described it, "Our army suffered. Influenza was rampant.... It was a grievous business having to listen every morning to the chiefs of staffs' recital of the number of influenza cases, and their complaints about the weakness of their troops if the English attacked again."[100]

By May the outbreak had largely run its course at Funston. During the first week of September influenza was not even included among the lists of diseases for concern by the Medical Department. Even in most of those cases in which troops either in Europe or the United States were stricken, the disease appeared relatively mild. British physicians were hesitant about even calling the outbreak influenza.

> A widespread epidemic of *grippe* in a mild form is at present invading Paris and a large part of France. It is probably an extension of the epidemic which has recently attacked the whole population of Spain, from the highest to the lowest. The cases are very numerous, but none so far of a grave nature. Complaint is made of headache, pain in the back, naso-pharyngeal catarrh which is persistent but rarely invades the lower respiratory tract. The temperature is moderately raised, dropping at the end of two days. The resulting indisposition, which seems to be spreading to the troops at the front, fortunately lasts at most a few days.... [In Belfast] the symptoms are mainly high temperature with a sudden onset, congested eyes, sore throat, and headache, with muscular pains, and in uncomplicated cases the duration of the fever is about three days. Occasionally there are sickness and bronchial catarrh, but pneumonia is not common. It resembles influenza in its sudden onset and its widespread epidemic character, but differs from former visitations of that disease in comparative absence of chest complications and post-febrile depression. The locus of the disease seems to lie in the throat, at the junction of the pharynx and larynx. Bacteriological examination so far has shown the presence of diplococcic, streptococci and pneumococci, but no specific organism [the author of *The Lancet* report noted].

After setting out three points in which the clinical symptoms differed from those generally ascribed to Pfeiffer's bacillus, the statement is made that in all the cases studied one organism predominated in direct smears from nasopharynx, throat and sputum ... resembling those of streptococcus. No colonies of *B. influenzae* [Pfeiffer's bacillus] were discovered. The authors are now working to establish Koch's third postulate [i.e., infecting a test animal] in relation to the organism.... We commend these series of investigations to the attention of bacteriologists in general. The loose application of the term "influenza" to a febrile infection in which Pfeiffer's bacillus cannot be demonstrated is, to say the least, unfortunate and we have always deprecated its use, suggesting as an alternative the term "catarrhal fever." Mild though the present form of the malady is in a vast majority of cases, its death-rate has not been insignificant, and its epidemiology is a matter of very pressing importance.[101]

Several observations may be made concerning the report which, despite the denial by the authors, is almost certainly a description of influenza. The authors place great weight on the inability to isolate Pfeiffer's bacillus, which we now are aware is the cause of secondary infections, but not of influenza itself. The authors did include a description of the infectious nature of the disease and the epidemiological implications. Other observers were even more convinced that what they were seeing was not influenza.

Is the present widely spread epidemic one of influenza or is it something new? In recent years influenza has been only endemic, hence there has been little opportunity to study its causative organism by modern laboratory methods. The last great epidemic began about 1889, and lasted until about 1892. It rapidly spread over the civilized world. The present epidemic is also widely spread. Several camps have experienced it. A few weeks ago the [censored] area began to have cases. It rapidly spread through bodies of men.... Clinically the disease simulates influenza. It is an acute febrile infectious condition of three or four days duration. The most striking symptoms are: Sudden onset with chills, severe headache with pain in cervical, dorsal and lumbar regions, also pain in limbs and general malaise.... It will be noted: (1) That the pyrexia is of shorter duration; (2) that the total course of the disease is shorter; (3) that the gastrointestinal symptoms are slighter than in the form of influenza due to the *Bacillus influenzae*, commonly called Pfeiffer's bacillus. It is also noteworthy that convalescence is rapid and that as yet we have noted no cases of relapse, recurrence, or of complications. It must be remembered that the cases under our observation were young healthy soldiers in prime physical condition.

The authors were further unable to isolate any specific species of bacteria, most notably in their minds, Pfeiffer's bacillus. Their conclusion was that the illness they were observing was not influenza. "The clinical course, though similar to that of influenza, is of very short duration, and there is, so far as we have observed, an absence of relapses, recurrence, or complications.... The organism of influenza—viz., *Bacillus influenzae*—was in all cases absent and there was present with no exception a Gram-positive diplococcus. [perhaps *Streptococcus*]."[102]

Once again the assumption that the etiological agent of influenza was a bacterium led to an incorrect conclusion. The conclusion, right or wrong, would prove to be moot in any case. Once the pandemic actually began, there was little that could be done either to slow or stop the spread, given the contingencies of war, or to provide palliative treatment for the symptoms with medicines available at the time.

But Funston was merely the starting point for soldiers moving east towards other cantonments, including Camp Devens, thirty-five miles northwest of Boston. The first week of September may have indeed been quiet, but using a cliché, it was the eye of the storm. Influenza reappeared in Camp Devens on September 8. From thirty to ninety hospital admissions each day for the next week the numbers quickly jumped to over five hundred on September 14. By the following week over 1,000 men were admitted each day. The numbers quickly overwhelmed the capacity of the camp hospital, which was constructed on the premise of a maximum of 1,200 patients. With the present outbreak nearly that number were being admitted daily, with nearly 20 percent of the camp reporting ill. The chief nurse at the hospital, Jane Malloy, calculated that "three miles of hospital corridors were lined on both sides with cots."[103] And it was not only the enlistees who were ill and dying; the medical staff were also among the patients. The Red Cross was requested to provide help for the doctors and nurses. Twelve nurses were sent; within days eight were ill. Two of the nurses died.[104]

The highly infectious nature of the outbreak was not unusual for influenza. But historically, most influenza epidemics were characterized by mortality highest among the most vulnerable: the very young or elderly. At Devens, as at Camp Funston, it was the healthiest and most robust who seemed most susceptible. The increasing number of deaths among these men was also highly unusual.

> These men start with what appears to be an ordinary attack of LaGrippe or Influenza, and when brought to the hospital they very rapidly develop the most vicious type of Pneumonia that has ever been seen. Two hours after admission they have the Mahogany spots over the cheek bones, and a few hours later you can begin to see the Cyanosis extending from their ears and spreading all over the face, until it is hard to distinguish the colored men from the white. It is only a few hours then until death comes.... One can stand it to see one, two or twenty men die, but to see these poor devils dropping like flies.... We have lost an outrageous number of nurses and doctors. It takes special trains to carry away the dead. For several days there were no coffins and the bodies piled up something fierce.[105]

The outbreak had been so explosive during September, seemingly appearing almost out of nowhere in a manner of days, and was so clearly unusually deadly that it could not be ignored. Gorgas soon ordered Victor Vaughan, William Welch, Rufus Cole and Frederick Russell to Camp Devens to observe first-hand the outbreak and determine what, if anything, could be done. The men had recently completed an inspection tour of army camps in the south hoping to head off any potential for an epidemic. What they had in mind was something more akin to the less virulent influenza which had appeared in the spring of that year.

After leaving Camp Macon in Georgia, Vaughan suggested they rest and tour the area around Asheville, North Carolina, where he had spent some time during inspection tours in 1898. A nearby estate had been constructed by the Vanderbilt family during the 1890s and had become a common retreat for the wealthy and famous. Vaughan and Welch in particular had grown closer in the years since their professional disputes in the 1890s. Vaughan and Welch had been dubbed the "gold dust twins" by another member of the Council of National Defense. "They were inseparable companions, and always appeared together at our meetings, sauntering down the long corridor, nodding friendly greetings to all the staff, their rotund bodies radiating cheer and good nature."[106] Welch and Vaughan had more than just their profession in common; each by this time in their lives weighed well over 250 pounds.

While visiting Asheville, Vaughan contracted a severe respiratory infection which delayed their return to Washington. Once they returned to the city on a Sunday morning they were met at the train by an escort who told them to immediately report to the surgeon general's office. Gorgas himself was in Europe, but when they arrived there Gorgas' deputy gave the order that they were to proceed immediately to Boston and Camp Devens.

After an eight hour train trip, the inspectors arrived at the camp to discover the situation was even worse than they had imagined. As Vaughan later described his feelings after the pandemic had run its course,

> I went to Camp Devens as soon as influenza was reported and the realization of the utter helplessness of man in attempts to control the spread of this disease depressed me beyond words. We are inclined to boast that the age of pestilence has passed, but, with a fair acquaintance with the history of epidemics I dare say that the world has never before known a pestilence more widespread, more intensive and appalling in its progress, or more destructive to life, than the epidemic of influenza which apparently came into being and grew in violence as the World War passed through its final stages. It seemed that Nature gathered together all he strength and demonstrated to man how puny and insignificant he and his forces are, with all his murderous machinery, in the destruction of his fellows.[107]

In the hospital it was obvious the outbreak was for the moment beyond control. The hospital had been built to hold twelve hundred patients though in an emergency that capacity could be doubled. By the time of Vaughan's and the others arrival, over six thousand patients had been admitted. Nearly half the nursing staff of two hundred were already ill. In Vaughan's words, "Hundreds of young stalwart men in the uniform of their country are coming into the wards of the hospital in groups of ten or more. They are placed on the cots until every bed is full and yet others crowd in. The faces wear a bluish cast; a distressing cough brings up the blood-stained sputum. In the morning the bodies are stacked about the morgue like cord wood."[108]

The autopsy room was like something out of a horror movie. As Rufus Cole later described the scene, "Owing to the rush and the great number of bodies coming into the morgue, they were placed on the floor without any order or system, and we had to step amongst them to get into the room where an autopsy was going on." Even Welch was shaken. "Welch opened the chest of the corpse of a young man, exposing his lungs. It was a terrible sight. When the chest was opened and the blue swollen lungs were removed and opened, and Dr. Welch saw the wet, foamy surfaces with little real consolidation, he turned. 'This must be some kind of new infection, or plague.'"[109] Other physicians as well noted the unique characteristics of the pathology, described as intense "congestion and hemorrhage." Lungs which normally would be "light and airy" were instead "sodden and distended, filled with a frothy, bloody liquid. Pathologists described how, when they moved a body to do an autopsy, 'a foamy, blood-stained liquid ran from the nose and mouth when the head was lowered.'"[110]

Victor Vaughan (Images for History of Medicine/National Library of Medicine).

At the beginning of the outbreak Camp Devens had nearly fifty thousand men present. By September 22 approximately 20 percent of the men were reportedly ill with influenza. The total dead would reach nearly 750 victims.

Following his inspection of the camp and autopsy room, Welch made several phone calls. In Boston he spoke with Dr. Simeon Burt Wolbach, professor of pathology and bacteriology at Harvard University, and described the illness. In Washington he spoke to Dr. Charles Richard, acting army surgeon general while Gorgas was still in Europe, requesting an expansion of existing hospital facilities, and possible quarantine of cases. And in New York, Welch spoke with Dr. Oswald Avery from the Rockefeller Institute hoping that Avery, a noted bacteriologist and expert in the study of the pneumococcus, would be able to isolate and identify the agent.[111] After leaving Camp Devens, Welch himself fell victim to influenza. Rather than entering a hospital upon his return to Hopkins, he chose to return to his rooms at Mt. Vernon, where he spent nearly six weeks recuperating.

On his own initiative, Richard took immediate action. On September 24 he issued a warning to the Medical Corps.

> Inasmuch as an epidemic wave of influenza is sweeping over certain parts of the United States and threatens to involve many of the camps and cantonments, it is important that the essential facts in regard to the nature and prevention of the disease be more generally understood.
>
> The disease now epidemic is believed to have been imported from Europe where it has been prevalent in various countries. Popularly called "Spanish influenza," there is nothing to indicate about it any departure from the influenza which has been prevalent in the United States from time to time for many years and was last seen here in army camps and cantonments in the spring of 1918.
>
> No disease which the army surgeon is likely to see in this war will tax more severely his judgment and initiative. It will be wrong on the one hand to propose such measures of prevention and treatment as will interfere unduly with the rapid training of the men, and on the other to make so light of the disease as to increase the sick rate from the more serious diseases which the influenza is associated.
>
> It is important that influenza be kept out of the camps, as far as practicable. To this end it must be recognized as a disease which is distinct and separate from the so-called "cold," bronchitis, laryngitis, coryza, or rhinitis, and "fever, type undetermined," which are continually with us and from time to time become prevalent. The influenza which is now epidemic is not a part of, or the cause of, or the consequence of any of these diseases. It is a specific infection with a characteristic symptom-complex.
>
> Upon the appearance of influenza in camp, special provision should be made for ample hospital accommodations for these patients. Owing to the great infectivity of the disease, sole reliance should not be placed upon cubicles and masks for isolation. It is needless to say that surgeons, nurses, and attendants should use every precaution against becoming infected themselves, and from carrying the virus to others.

> There are few diseases so infectious as influenza. The virus is contained in the discharges of the nose and mouth, and is given off in the acts of sneezing, coughing, speaking. The hands and whatever else may become contaminated by the discharges, can carry the virus and produce new cases. It is probable that patients become foci of infection before the actual symptoms develop and remain so after the active symptoms subside. Influenza is a disease which is often produced by carriers.[112]

Richard had a clear understanding of the nature of the disease and its method of spread. Though he did make reference to the "virus" of the disease, it should be kept in mind that this was a reference to the putative etiological agent, not in the literal sense as we know the agent today. Richard also recommended to Army Chief of Staff Peyton March that in light of the spread of the disease through army camps, draft calls should be temporarily halted and that movement of men between camps or onto troop ships heading to Europe should be reduced. But the exigencies of war rendered these recommendations impractical, with the expected consequence. Men on troop ships became infected with the more virulent strain and the disease was brought to Europe.

Avery's attempt at identifying the etiological agent provided the single element of dark humor to the tragic background. Following the phone call from Welch, Avery traveled to Boston and from there to Camp Devens. While the general assumption in the minds of many of the physicians was that Pfeiffer's bacillus, *Bacillus influenzae*, was the etiological agent of the disease, Avery preferred to withhold judgment until he could complete his own investigation. The first step in attempting to identify a bacterium is to carry out what is referred to as a Gram stain. Developed by Christian Gram in the nineteenth century, the test involves staining of autopsy specimens with a purple stain, crystal violet, followed by subsequent steps which include an alcohol decolorization procedure. The final step utilizes safranin, a red stain. Depending upon the cell structure, the cells will stain either blue (Gram-positive) or red (Gram-negative), as is the case with Pfeiffer's bacillus. The alcohol step is critical.

Avery was unable to detect the presence of the bacillus in any of the patients from whom he obtained specimens. At some point he ran out of alcohol for decolorization and obtained a fresh supply. Now the Gram stain worked properly and he was able to detect the presence of the bacillus. He discovered that the alcohol he had been using was actually water; soldiers had drunk the original supply and replaced it with the water. With the proper reagents, Avery detected the bacillus in twenty-two of the thirty autopsy specimens he analyzed. For some, this was sufficient to establish the etiological agent of the outbreak as Pfeiffer's bacillus. As Wolbach described the autopsy findings,

> Because of their small size the influenza bacilli are easily recognized. It is interesting to note that in some early cases the bacilli were found not only in the bronchial exudate, but in the submucosa of bronchi and in the alveolar walls of the lung.... The influenza bacillus was found in pure culture in one or more lobes in nine of the twenty-three cases from which cultures were made. In sections of lung from cases in which no cultures were made, influenza bacilli were found apparently pure in two cases, and mixed with other organisms in one case.... In a number of cases in which influenza bacilli were not found in the lungs by culture, they were found in cultures from the sinuses of the skull or from the middle ear.
> The pathology of the lungs indicates clearly that we are dealing with a specific infection with a distinctive pathology in its early stages. The occurrence of *Bacillus influenzae* in pure cultures in the early stages is a fact of importance in the consideration of the etiology of influenza and I believe firmly establishes the existence of a *B. influenzae* pneumonia.

Wolbach considered other bacilli isolated from lung tissues in these patients to merely represent secondary invaders. That Pfeiffer's bacillus might represent the same form of secondary infection was not strongly considered.[113]

Wolbach was not the only person close to the investigation to be fooled. George Soper as well worked on the assumption that the etiological agent was a bacterium. Soper also recognized that sometime between the first outbreaks in that spring and the pandemic which began in September that the agent—mistaken by Soper like others as a bacterium—had mutated into a much more virulent form. "The causative agent is believed to be the bacillus of Pfeiffer; the means of transfer; the air and objects recently contaminated by the buccal and nasal secretions of those who harbor the virus. It is a fundamental assumption that influenza is produced when, and only when, material from the mouth or nose of infected persons gets into the mouth or nose of someone who is susceptible. As is plainly recognized in respect to intestinal infections, the hand probably plays an important part in the transmission of influenza. Coughing and sneezing help greatly to spread infection." When discussing the change in virulence of the agent, Soper further wrote,

> It has long been known that interchanges of bacteria occur commonly from mouth to mouth under ordinary conditions of social intercourse. Most of the organisms are harmless under normal conditions of health. That their virulence is sometimes increased, sometimes reduced, according to circumstances, appears to be certain. But what the circumstances are which raise or lower the virulence is conjectural. The Pfeiffer bacillus is no stranger to America; it was believed to be present in many healthy persons before the present pandemic. To account for the pandemic it has been suggested that something must have happened to

increase its virulence or a new and more active strain has appeared, or the susceptibility of those attacked has become greater.

The belief that immunity is conferred by an attack is partly confirmed by the observation that in Europe and America a preponderance of persons who have suffered in the present pandemic are relatively young persons, few of whom could have experienced the disease during the pandemic of 1889-90.

The weather has always been supposed to exert an influence upon influenza—the very name is derived from the effect which extraterrestrial conditions were supposed to exert upon it. But although there has been a great deal of study of this subtle matter, little is known concerning it.

The epidemics which occurred in the spring of 1918 were like those which are taking place now, except that the disease was milder and there was less pneumonia. Until recently the influenza reported from Europe was of this mild type. It seems to have been as infectious as it is now. Reports coming from all parts of Europe indicate that the percentage of persons attacked was about the same as that at present.

Something seems to have occurred during the summer greatly to increase the virulence of the disease. During July and August a number of vessels plying between Europe and America experienced intense outbreaks of influenza, accompanied by very fatal pneumonia. That cases of the disease were being brought into the country in this manner was stated in the daily press and in official reports in July.

Soper then proceeded to compare the current outbreak of influenza to those in the past. "It is impossible to say how severe were some of those which are recorded in history for the reason that statistical data concerning them is meager and imperfect. It is said that in 1889-90 no less than 25 per cent of the populations was attacked in London; 33 in Antwerp; 39 in Massachusetts, and in Paris, 50. In 1832-33 about 40 per cent of the population of Paris is believed to have been affected. In 1872, three quarters of the population of London and some German cities were supposed to have suffered."[114]

Vaughan himself suggested that Pfeiffer's bacillus might be the actual etiological agent, as on September 27 he, Welch and Cole indicated such to the surgeon general.[115] But by that December he was not as certain.

> Pandemic influenza is an acute infectious disease, of unknown causation, characterized by a marked leukopenia [loss of white blood cells] which removes the normal barriers against infection and exposes the body to invasion by whatever pathogenic organisms may gain access to it. It may cause death without the aid of other viruses, but in the majority of instances its deleterious effects are intensified by the presence and activity of other pathogenic organisms.... The pathogenic organisms which most frequently take advantage of the defenseless state of the body induced by the influenza virus are those which are capable of growth and multiplication in the pulmonary tissue, such as the different types of pneumococcus, streptococcus, Friedlander's organisms [*Klebsiella*], and possibly some varieties of staphylococcus and Pfeiffer's bacillus.... Influenza is not transmitted through the air for long distances, nor is it due to weather, nor to meteor-

ological conditions. It is transferred from man to man either directly or indirectly; either by direct contact or by indirect contact.[116]

For perspective on the extent of the outbreak, Soper provided the incidence of reported cases on influenza and pneumonia present in the twenty largest camps and cantonments during September and October:

	Total cases influenza	Total cases pneumonia	Total deaths pneumonia	Percent attacked influenza	Percent deaths pneumonia
Sept. 12–18	45,789	7,671	2,862	20.6	37.3
Sept. 22–24	42,267	7,399	2,591	21.2	35
Sept. 29-Oct. 1	32,932	6,818	2,280	21.8	33.6
Oct. 3–11	17,307	1,236	210	22.8	17.8

Twenty camps in chronological order of attack (adapted from Soper[117]).

As presented in the table, the number of reported cases among troops during the four-week period with the highest incidence approached 140,000 men, among whom there were approximately 7,600 deaths. The total level of morbidity and mortality before the disease ran its course that fall was staggering. By a conservative estimate using numbers from the War Department, over one million men were stricken with influenza, approximately 26 percent of all soldiers. Of these almost 30,000 died, most before they even had the "opportunity" to land in France. Nearly half the deaths among American troops were directly or indirectly the result of influenza. There is no reason to think numbers were not comparable among German troops.[118] Numbers in the civilian populations around the world are at best estimates. Conservative figures suggest 500 million persons were clinically ill with influenza, roughly one-third the world's population. Of these between fifty and seventy-five million persons died, nearly 3 percent of the world's population. In the United States an estimated 675,000 persons died, 195,000 in October alone, out of a total population of some 104 million; estimates are that 25 percent of the population became ill during the pandemic.

As if the disease wished to demonstrate one final gasp, a third wave of influenza began in the winter months of 1919. Soldiers who had survived combat in France were now succumbing to an illness which had for a brief time appeared to have begun dying out.

If the same or similar cross-reacting strains of the influenza virus were responsible for the three waves, including that in the fall which was the most deadly, it is reasonable to ask whether exposure to the less virulent form of the virus in early 1918 provided any measure of protection in the second and third waves. The answer appears to be yes. While attempting to include the caveat that while estimates are largely based upon reported illness among soldiers, it is important to recognize that men may have been exposed while still civilians. The numbers suggest that exposure during the first wave of influenza provided between 35 percent and 94 percent protection against illness, and between 56 percent and 89 percent protection against death.[119]

As an example of a pandemic disease, influenza stood apart. This is not to say, however, that lessons in disease prevention had not been learned from previous conflicts. As Vaughan pointed out shortly after the war, typhoid fever was no longer a significant factor in disease mortality. During the Spanish-American War the annual death rate in army camps from typhoid fever was 879 per 100,000. During 1918 it was 1.3, close to non-existent. During the previous war every regiment in the army included cases of typhoid, in some instances as many as 20 percent or more of the recruits. Few regiments reported typhoid in 1918.[120]

It was not only the influenza pandemic with which Vaughan became associated during the war. Even prior to the entrance to the war by the United States, psychologists had expressed concern with the psychological impact of removing civilians from their familiar home lives and subjecting them to extensive military training, let alone the horrors of war itself. Among these issues addressed by psychologists were included methods for accurate psychological examination, recreation and amusement, physical challenges such as eyesight, incapacity due to what was then referred to as shell-shock, emotional instability and fear.[121] Vaughan was among those who developed a series of written and oral psychological examinations in order to help determine mental fitness for life in the army and in the endurance of combat conditions.

Chapter 10

Retirement

While Vaughan's official duties as dean of the medical school came to an end in 1921, the man never truly retired until declining health left him no choice. Time had finally taken its toll, and at the age of seventy Vaughan no longer had the stamina to truly carry out his job to the best of his ability. As a result, Vaughan's resignation from his position as dean was presented to the university board of regents during the March meeting that year. Along with the resignation submitted by Vaughan was a resolution presented by Regent Walter Sawyer which the faculty of the medical school had in part provided on his behalf:

> *Resolved,* That with the acceptance of Dr. Victor C. Vaughan's resignation as Dean of the Medical School and Professor of Hygiene and Physiological Chemistry, the Regents give expression to their deep appreciation of his long and distinguished service to the University and to medical education, and of his contributions to science.
>
> The world recognition of him as an administrator and a worker in the field of research has been a large factor in placing of the Medical School in a position of first importance and standing.
>
> Honored by membership in international and national scientific and social welfare bodies, and having held a high office in many of them, he has had a rare opportunity for influence and vision in race progress and betterment. Responding to his country's call, he has in two wars acted as a valued advisor and investigator, and by his aid in the determination of the causes of disease and the application of corrective measures, helped to bring about the greatly reduced morbidity and mortality. Few men can measure up to his attainments and accomplishments, and the international estimate of them.
>
> *Resolved,* further, That the Board of Regents wishes for him a happy relief from his arduous labors, and his continued interest in the Medical School, for the success of which he has so long been responsible."[1]

The regents also allowed Vaughan continued use of his laboratory for any additional scientific research he might be interested in carrying out. In addition, the regents authorized the president of the university, Marion Burton, to request the Carnegie Foundation for the Advancement of Teaching provide a retiring allowance for Vaughan. Following the motion of the board of regents member James Murfin, the resolution and requests were approved unanimously, and on June 30, 1921, the thirty-year tenure of Victor Vaughan as dean of the Medical School came to an end.

In September of that year, Vaughan and his wife made a temporary move to Chevy Chase, Maryland, where he had accepted an offer to serve as chairman of the Division of Medical Sciences of the National Research Council, a position he held until October 1922. The National Research Council had been established by President Wilson during the recent war, and Vaughan had been a member (see Chapter 8). The NRC became a permanent organization in 1919. The role of the council by 1921 had evolved to that of a liaison between the government and private sector researchers, and overseeing that research. Among the accomplishments of the division during Vaughan's tenure as chairman was the awarding of a $500,000 grant from the Rockefeller Foundation for fellowships to train medical personnel interested in teaching or careers in research; the period of the fellowships was to run from January 1, 1922, to December 31, 1926. Among the more prominent recipients was Alfred Mirsky, who went on to become one of the pioneers in the future field of molecular biology.[2]

While residing just outside of Washington during these early years of retirement from the university, Vaughan took advantage of the wealth of resources available within Washington archives as well as those in Michigan to produce a two volume work, *Epidemiology and Public Health* (1922). The work began with an abbreviated summary of his life and medical career, foreshadowing the much more detailed description provided in his autobiography completed several years later.

Vaughan's life had been concurrent with the growing understanding of the infectious nature of disease. Until the mid nineteenth century the prevailing theory of most diseases was their relationships to miasmas, the odors or emanations from the soil which, when inhaled, somehow resulted in the development of illness. As an illustration of how this theory was applied, Vaughan used the example of Max von Pettenkofer and typhoid fever. Pettenkofer was among the most important chemists and hygienists, arguably the father of modern hygiene, in Germany through much of the nineteenth century. But it was Pettenkofer's theory that typhoid fever, as well as cholera, were the products of the "ripening" or putrefaction of fecal material. As denoted by Vaughan in *Epidemiology*, Pettenkofer used the example of the Munich water supply to bolster his argument. Munich was considered a "hotbed" of typhoid, and a city overrun with fecal contamination of the city water supplies; "fresh" fecal material, upon ripening, would develop into the source of typhoid was the basis for Pettenkofer's argument. In response to Pettenkofer's theory that the source of the disease was the emanations from the polluted water, the city began to use fresh mountain water. Not surprisingly, typhoid was largely eradicated (in addition to much of the odor). For Pettenkofer, this was vindication of his theory. However, once the typhoid bacillus was isolated, the results which had taken place in Munich were explained in accordance with the germ theory of disease.[3] Vaughan applied the same logic during his investigation of typhoid outbreaks among troops during the Spanish-American War.

Pettenkofer later used the same argument of soil emanations in support of his theory explaining the origin of cholera. The emanations from putrefying soil activated the cholera agent; in the absence of either soil or bacteria, cholera could not develop. Pettenkofer followed this idea to an extreme, swallowing a beaker of the cholera bacillus to illustrate his point; whatever the reason, he suffered no significant ill effects.

The use of epidemiology in explaining the source and development of disease was still relatively new at the time Vaughan produced this work. If one assumes the work of John Snow in 1849 and 1854 in explaining the sources of cholera outbreaks in London during those years represented the first modern application of the science, the historian could argue that much of the knowledge of epidemic disease was developed in the lifetimes of some still alive at the time Vaughan produced this compendium. Vaughan's justification for the emphasis on understanding the spread of disease is as true today as it was in the 1920s.

> Works of epidemiology have been devoted almost exclusively to histories of past epidemics. This is essentially true and proper, because there is no way we can get information except through the study of things already accomplished. Now, however, the field of epidemiology has broadened and it includes everything that pertains to epidemic diseases. The epidemiologist must be a student of etiology, symptomatology and pathology. The exercise of his function should not await the wide development of disease. Every case of an infectious disease is potentially the seed from which many cases are to develop. The bacteriologist need not know much about pathology or symptomatology in order to be useful and learned in his specialty. The pathologist may make valuable contributions to medical knowledge without himself having any wide acquaintance with etiology or symptomatology, but a knowledge of all these branches is required of the epidemiologist.[4]

Vaughan also used the opportunity afforded by his writing to again explain his own theories as to the mechanism of many diseases: protein poisoning. After reviewing much of his scientific work, Vaughan provided additional examples to explain his ideas. "We venture to suggest that our experience in attempting to immunize animals with the protein poison may be the basis of a scientific explanation of the frequently observed fact that in epidemics of respiratory disease in armies the recent recruit and the man from a rural community fall more ready victims to the epidemic than do his comrades among seasoned soldiers and among those who have lived in crowded cities. The man who has inhaled for a long time and quite constantly a large number of bacteria whose proteins have been split in his body, setting the poisonous group free, has acquired a tolerance for the poison or an increased resistance to infection from any bacteria taken in through the respiratory organs."[5]

Vaughan continued to apply his theories concerning the splitting of proteins as being the underlying cause of the pathology associated with disease, whether sensitization to protein or in extreme cases, anaphylaxis, a term Vaughan preferred to avoid, or in development of bacterial immunity. His arguments were largely reiterations of those expressed earlier in his career and in earlier publications (see Chapter 7).

> Bacteria and protozoa are living, labile proteins, while egg-white, casein, serum albumin, etc., are stabile proteins. The proteins of one group are in an active, while those of the other are in a resting, state; but both are essentially poisons made up of an acid or poisonous chemical nucleus, and basic, nonpoisonous groups. Bacterial immunity and protein sensitization, apparently antipodal, are in reality the same, and each consists in developing in the animal body the capability of splitting up specific proteins. If the living protein be split up before it has time to multiply sufficiently to furnish a fatal quantity of the poison, the animal lives and we say it has been immunized. If the stabile protein be introduced into the animal it leads to the development of a specific proteolytic ferment, and if enough of it to supply a fatal dose be reinjected after this function has been developed, the animal dies.... We are fully aware of the fact that animals, especially guinea pigs, may be killed with all the symptoms of anaphylactic shock by primary injection of various substances, such as fine suspensions of agar, dilute incubated blood serum and tissue extracts, but so far as we can see these have nothing to do with anaphylaxis or protein sensitization as applied to the infectious diseases. An extract of an organ from a fellow of its own species injected intravenously or intracardially into a guinea pig will kill it with all the symptoms of anaphylactic shock, but this has nothing to do with protein sensitization and certainly has no bearing on the relation between anaphylaxis and the infectious diseases. As we have demonstrated, the guinea pig possesses a high degree of immunity to the prodigiosus [likely *Serratia*], but when the prodigiosus is grown artificially and thrown in sufficient dose into the abdominal cavity of the guinea pig the animal quickly dies. In our opinion, the animal dies because the secretions of the body cells speedily break up the prodigiosus protein, setting free in a short time, a fatal dose of the protein poison. The more highly immune an ani-

mal is to a given bacterium the smaller the dose of the cellular protein of that bacterium necessary to kill that animal.[6]

Vaughan summed up his introduction by asking what differentiates a saprophytic organism, one which is not a pathogen, from those which exhibit pathogenicity. He believed either that saprophytic organisms feed only on dead matter rather than on the fluid or materials from the animal, or that the body's ferments digest the saprophytic cells. On the other hand, parasitic bacteria are capable of survival in the host as well as digesting fluids or tissue materials in that host. A bacterium may also be nonpathogenic if it resides in one area of the body, becoming pathogenic when it travels to a different region of the body. Vaughan used such examples as the diphtheria bacillus which he observed in saliva, the typhoid bacillus in the gall bladder and the meningococcus which grows in the nasopharynx to bolster this argument.[7]

Throughout *Epidemiology* Vaughan continued with a common theme: the epidemiology of disease. However, not all Vaughan's analyses were those of infectious disease. The first examples analyzed in this work were those of allergies: hay fever and even forms of asthma. Even here though, Vaughan continued to argue the underlying mechanism was that of a protein poisoning.

> When a foreign protein is carried into the blood and is distributed throughout the body it penetrates the body cells and leads these cells to produce a new specific ferment which digests the invading or foreign protein. This new function developed by the body cells may be permanent or it may be temporary. As long as it exists, whenever the cell subsequently comes in contact with this foreign protein the cell pours out its proteolytic ferment and splits up the foreign protein with the production of the protein poison. The symptoms of pollinosis are those induced by the reinjection of a given protein to which the animal has been sensitized. Some years ago [John] Auer and [Paul] Lewis in studying the phenomenon of protein sensitization suggested that it might lead to asthma. This has turned out to be true and has been confirmed by everyone who has worked on the subject. We do not mean to say that every form of asthma is due to protein sensitization, but we do mean to say that the asthma of hay fever is a symptom of protein poisoning.[8]

Though Vaughan was not aware of the molecular mechanisms which are the reasons for the allergic response, the knowledge of characteristics of hay fever allergies differs little from what we currently know.

> The intensity of the symptoms borne by a victim of pollinosis depends upon the abundance of the pollen in the air and the readiness with which the unbroken protein is absorbed and taken into the circulation. Even in the same neighborhood, in the different seasons or on different days during the same season, the number of pollen in the air varies greatly. Rain carries the pollen grains to the earth and at the same time, in part at least, washes out the proteins and these

doubtlessly are largely absorbed in the earth. The hay fever sufferer may take advantage of the fact that the pollen proteins are readily soluble in water. Wet curtains hung at the windows will catch many pollen grains and dissolve out their protein constituents. Washed pollen grains do not sensitize. The sensitizer lies wholly in the soluble part of the grain.[9]

Vaughan was also aware of the medical procedures available for desensitization of the sufferer and which in some cases could provide at least temporary relief from the allergy. "In the present state of our knowledge there is in reality no such thing as the [permanent] desensitization of a sensitized animal. Such an animal owes its sensitization to rather radical changes in the molecular construction of the proteins of its cells.... Some proteins in sensitizing cause deep and permanent changes in the molecular structure of body cells, while others cause only slight and temporary effects. In either case science up to the present time knows no method by means of which a cell or a protein molecule which has developed a new function can be made to lose it."[10]

Vaughan also described how some hay fever sufferers could obtain some relief from the allergy. The pollen protein was dissolved in a solution of alcohol and increasing amounts of a diluted solution were injected under the skin of the patient. In many patients the procedure allowed for a desensitization which lasted through one season; the procedure would have to be repeated the following year. A similar procedure is still utilized today for relief from certain types of allergies. The molecular mechanisms associated with either the allergy itself or the basis for desensitization were not known during Vaughan's lifetime. As we are now aware, allergies such as those of hay fever are associated with a specific type of antibody, IgE; a similar antibody is associated with anaphylaxis. Using a simplistic explanation, desensitization may occur when, after the injection of a small quantity of allergen, a second type of antibody is produced which can eliminate the protein allergen before the body has an opportunity to react.

Each subsequent chapter in *Epidemiology* dealt with a specific infectious disease, ranging from respiratory illnesses such as the common cold or influenza to bacterial infections such as tuberculosis or typhoid. Since the thesis of the work is that of epidemiology, each chapter dealt with the history of the disease as well as its mechanisms of transmission. The material was current with the time during which it was written; for example, the chapter on the common cold described the isolation and identification of the virus which is the etiological agent, a determination made only a few years prior to the publication of the work.

The next three years were spent by the Vaughans in traveling, though he still found the time in 1923 while in Chicago to participate in the startup of a new journal overseen by the American Medical Association (AMA): *Hygeia*. Named for the daughter of the Greek and Roman god of medicine, the journal was established within the auspices of the Bureau of Health and Public Instruction of the AMA to promote medical instruction and information among the general public. It ran until 1949 when, under the direction of a new editor, it continued for several decades under the title of *Today's Health*.

In September 1925, Vaughan again accepted an offer to serve as chairman of the Medical Division of the National Research Council. One of the pressing problems which soon came before the members of the medical committee was that of explaining and solving the growing number of homicides which were occurring in the country. Vaughan was in favor of carrying out a scientific study of the problem, and believed that as a perceived medical issue, it fell within the purview of the NRC to address. A committee was established, and Vaughan asked Ludvig Hektoen, a Chicago pathologist and noted advocate on issues dealing with public health, to head that committee. At the time of his appointment to the committee, Hektoen was already the director of laboratories at the McCormick Institute for Infectious Disease in Chicago and was serving as chair of the Division of Sciences within the NRC.

It was Hektoen's opinion that medical forensics, as the science later became known, was in a poor state within the country, with politics and a surfeit of legal experts underlying the problem. The office of medical examiner, now a standard position associated with the coroner's office in most cities, had only been recently created (1925), and with relatively few students with expertise in the subject being trained in the medical schools, the problem would only become worse in time. Vaughan had initially hoped that the Rockefeller Foundation would help direct the study, but Alexander Flexner replied that while the foundation would help in the form of funding, it would not participate in the actual study. Hektoen's committee eventually produced a 101 page report, released in 1928, which in addition to castigating the incompetence exhibited by coroners in many cities, held the examiner program which had been established in New York as a model to be followed by the rest of the country.[11]

Despite his busy schedule during these years of "retirement" Vaughan still took the time to produce his extensive autobiography, *A Doctor's Memories* (1926). The book has continued to serve as a compendium of his life, both professional and personal, with extensive descriptions of many of his experiences. Though little is related about his family life, particularly as it applied to his five sons, the story as provided by Vaughan described the many friends he developed among colleagues both national and international. The importance of his work as a member of the Typhoid Commission in particular is a highlight of that portion of the book.

Though much of Vaughan's work was in the field of bacteriology, his underlying working principle was always that of the sister fields of chemistry and biochemistry. His theories on the generation of protein poisons, whether as ptomaines or leucomaines depending upon their origin, remained largely extant even into his years of retirement. Those years also corresponded with the discovery of viruses and their relationship to some human diseases. The observation of viral particles would require the development of the electron microscope, a technology still in the future at the time. But it was clear to Vaughan that these filterable agents were capable of reproduction and exhibited at least some of the characteristics of being "alive." Given Vaughan's interests in biochemistry and mechanisms of bacterial pathogenicity, it was only a small step for him to develop his own theories about concepts such as "alive" vs. "not alive," and even the origin of life.

In 1927 Vaughan was invited to speak before the assembly of the 73rd meeting of the American Chemical Society, held in Richmond, Virginia, the week of April 11. His presentation was scheduled for the third day of the meeting, Wednesday. However, due to illness, Vaughan was unable to attend that day and the work was presented by his son, Dr. Warren Vaughan, in front of some 1,300 attendees.[12]

The theme of the presentation centered on the origin of life, including the differences between inert material and that which could be considered alive; characteristics of what constitutes alive were a significant portion of the thesis. It is important to keep in mind the context of bacteriology during this period if one reads Vaughan's words. DNA, while having been discovered some sixty years earlier, had yet to be shown to serve as the genetic material in the cell. Bacterial viruses, bacteriophage, had only recently been discovered by Felix d'Herelle (and others), and some human diseases had been shown to result from viral infections. But whether viruses were actually alive was subject to interpretation; Vaughan believed they were.

In Vaughan's discussion of what he believed to be the origin of life he began at the simplest levels, that of the atom and its formation into molecules such as proteins. Vaughan did not consider proteins to be alive. He did speculate that in the presence of some form of energy, a stimulus, proteins may become a component of cells which are themselves alive. Living matter must respond to that stimulus, and using some form of metabolism, be capable of receiving or releasing energy. A second characteristic of life is that of reproduction. On the basis of these two attributes, bacteria would be considered by Vaughan to be alive.

So where did the energy necessary for life originate? The mechanisms of photosynthesis were poorly understood at the time, but Vaughan was correct in his assumption that energy derived from the sun was the source of energy in cells. Whether the sun provided the energy for conversion of inanimate matter into life was speculation; Vaughan believed it did. In support of this idea Vaughan drew on experiments carried out by Benjamin Moore and T.A. Webster, who demonstrated some years before that formaldehyde could be formed by exposure of CO_2 to ultraviolet light, and that formaldehyde could in turn polymerize into sugars.[13] It was further likely that inorganic materials could also be converted into amino acids in a similar manner. He acknowledged that how this would subsequently evolve into the living cell was unknown.

Victor Vaughan (Bentley Historical Library, University of Michigan [Victor Vaughan file]).

So what, in Vaughan's opinion, was the smallest and simplest living substance? Based on characteristics which a living substance should exhibit, he believed it to be the bacteriophage. When placed in a "heterologous" medium, a solution containing bacteria, phage could transform the substance into progeny bacteriophage. Phage were obviously capable of reproduction; they could adapt—mutate and evolve—though Vaughan did not use those terms. Phage were antigenic and composed of protein. He accepted d'Herelle's generic name of protobe (first life), representing the simplest form of life today. The nature of human viruses—Vaughan used the example of smallpox—was unknown at the time, but Vaughan did not rule out the possibility of their being protobes of the same nature as bacteriophage. While our current definition of "alive" generally does not include the category of viruses, given the criteria utilized by Vaughan and the state of knowledge for the time, his argument certainly could be justified.

Vaughan completed his ideas in "Concept of Origin and Development" by speculating on the role of adaptation both of bacteria as well as that of higher forms of life in changing environments. Emphasis was placed on the alterations of proteins; it bears repeating that the role of DNA remained unknown in 1927. Vaughan understood Darwinism (despite his use and application of the phrase "inheritance of altered or acquired characteristics"), though he avoided the use of the term.

> The constancy of bacterial types and indeed of all living substances depends upon a relatively unchanging environment. In the lower forms of life environment has a very definite influence upon the characteristics of life. Furthermore, alterations in the structure of the protein molecule resultant on environmental changes may be and are inherited. A microorganism living in a milieu in which the pabulum is readily assimilated and transformed into homologous proteins will thrive. If, on the other hand, the environment is one in which the available food material is of widely different constitution from that of the living substance, continued existence will depend upon the ability of the microorganism to elaborate a ferment capable of disintegrating the foreign protein or protein-like substance into its constituent amino acids so that they may be available for assimilation. If some of these amino acids are deficient in quantity for the particular living substance, continued existence will now depend upon the ability of the living structure to adapt itself to this deficiency. If such an adaptation is made, there will result a change in the make-up of the living protein molecule.... The presence and availability of new or different amino-acids or similar protein radicals will ultimately determine an alteration in the constitution of the living protein molecule. If this alteration in environment is permanent the altered constitution of the living molecule will likewise become permanent and will remain so as long as the environment is relatively the same.... The persistence of new species so formed is dependent upon the permanency of the environmental changes.[14]

To illustrate his argument, Vaughan used the example of streptococci which lost their pathogenic potential for an animal once they were passaged through a different organism; in addition to loss of virulence, the changes included alterations in its antigenic structure.

Vaughan summarized his view of life, as befitted a chemist, with an emphasis on the importance of the environment on heredity. "Morphologists stress the stability of germ plasm, but some of them do admit that certain poisons such as alcohol, lead, mercury and syphilis may deleteriously affect the reproductive cells.... A boy and a girl, born of healthy parents and raised to maturity under normal conditions may migrate into a goiterous district and after acquiring goiters may marry. Their children may [exhibit developmental problems]. In this case it is the absence ... of iodine in the food and drink which leads to this deterioration.... The claim that the reproductive cells are not influenced by the somatic cells is one which I believe to be unwarranted."[15]

"Concept" was to be the last of Vaughan's hundreds of publications. In October and November 1926, the Third Pan-Pacific Science Congress was scheduled to meet in Tokyo, Japan. Among the topics which the congress planned to address was that of organizing a permanent association of scientific institutions in the Pacific area, certainly a goal which Vaughan would find admirable. Approximately forty members of the American scientific community planned to attend, among whom was Victor Vaughan (accompanied by his wife). While traveling in the Orient, Vaughan took advantage of the chance to visit other areas, including China and the Philippines. Vaughan became ill upon returning to the United States, suffering an apoplectic seizure (likely a mild stroke) from which he could not completely recover. In declining health for the remainder of his life, Vaughan died November 21, 1929, in Richmond, Virginia, from a sudden heart attack.

Chapter Notes

Preface

1. Victor Vaughan, "A Chemical Concept of the Origin and Development of Life: A Preliminary Presentation," *Chemical Reviews* (July 1927) 4 (2): 167–188.
2. Horace Davenport, *Not Just Any Medical School: The Science, Practice and Teaching of Medicine at the University of Michigan, 1850–1941* (Ann Arbor: University of Michigan Press, 1999). Davenport (1912–2005) was William Beaumont Professor Emeritus of Physiology at the University of Michigan.
3. Victor C. Vaughan, *A Doctor's Memories* (Indianapolis: Bobbs-Merrill, 1926).
4. Frederick G. Novy, "Victor Clarence Vaughan," *Science* (December 27, 1929) 70 (1826): 624–626; *The Journal of Laboratory and Clinical Medicine* (June 1930) 25 (9): 817–942.

Chapter 1

1. William N. Hubbard and Nicholas H. Steneck, *The Origins of Michigan's Leadership in the Health Sciences* (Ann Arbor, MI: Historical Center for the Health Sciences, 1995), pp. 36–37.
2. Ibid., p. 36.
3. Ibid., p. 37.
4. Ibid.
5. University of Michigan, Board of Regents, *Regents Proceedings: 1837–1864* (Ann Arbor: University of Michigan, August 1847), 370.
6. Ibid. (January 1848), 391–392.
7. Hubbard, *op. cit.*, p. 38.
8. Ibid., p. 39.
9. Howard Markel, "The University of Michigan Medical School, 1850–2000: An Example Worthy of Imitation," *The Journal of the American Medical Association* (February 16, 2000) 283 (7): 915–920.
10. Horace Davenport, *Not Just Any Medical School: The Science, Practice and Teaching of Medicine at the University of Michigan, 1850–1941* (Ann Arbor: University of Michigan Press, 1999), p. ix.
11. Ibid., p. 1.
12. Ibid.
13. Ibid., p. 4.
14. Harvey Cushing, *The Life of Sir William Osler* (Hamburg, Germany: Severus Verlag, 2010), p. 307. Also cited in Gerald Imber, *Genius on the Edge: The Bizarre Double Life of Dr. William Stewart Halsted* (New York: Kaplan, 2010), pp. 106–107.
15. http://elane.stanford.edu/wilson/html/chap22/chap22-sect3.html.
16. Ibid.
17. Davenport, p. 17.
18. *Proceedings of the University of Michigan Board of Regents: 1864–1870* (June 1869), p. 338.
19. Janet Tarolli, "First Ladies at Michigan in Medicine," *Medicine at Michigan* (Fall 2000) 2(3); http://www.medicineatmichigan.org/magazine/2000/fall/women/.
20. Markel, *op. cit.*
21. Tarolli, *op. cit.*
22. Ibid.
23. Love Palmer and Henry Frieze, *Memorial of Alonzo Benjamin Palmer* (Cambridge, Mass: Riverside Press, 1890), pp. 173–177.

24. Davenport, *op. cit.*
25. Ibid.
26. *Proceedings of the University of Michigan Board of Regents* (October 1891), p. 552, http://quod.lib.umich.edu/u/umregproc/ACW7513.1886.001/562.

Chapter 2

1. Victor C. Vaughan, *A Doctor's Memories* (Indianapolis: Bobbs-Merrill, 1926).
2. Vaughan's grandfather, Colonel William Dameron, had commanded local militia in the 1832 Black Hawk War. Vaughan, *op. cit.*, p. 46. For a summary of the role played by militia see Richard Adler, *Cholera in Detroit: A History* (Jefferson, NC: McFarland, 2013).
3. Vaughan, p. 96. Harrington later became president of the University of Washington. In 1899 he mysteriously disappeared, reappearing nine years later as an inmate in the New Jersey State Hospital in Morristown, New Jersey. According to his wife, Harrington's condition was the result of having been struck by lightning.
4. Ibid., p. 98.
5. Ibid., pp. 99–100.
6. Silas Hamilton Douglas, Samuel Townsend Douglas, Ashley Pond, University of Michigan, Board of Regents, *The Regents of the University of Michigan vs. Preston B. Rose, Silas H. Douglas, Appellant,* State of Michigan, Supreme Court, in Chancery, 1881, p. 2.
7. Davenport, p. 6.
8. Vaughan, p. 101.

Chapter 3

1. University of Michigan, Board of Regents. *Proceedings of the Board of Regents (1876–1881)* (June 26, 1877), p. 136.
2. Ibid., p. 408.
3. *Proceedings (1881–1885)* (June 27, 1883), p. 360.
4. *Twelfth Annual Report of the Secretary of the State Board of Health of the State of Michigan for the Fiscal Year Ending September 30, 1884* (Lansing, MI: W.S. George, 1885), January 8, 1884, p. xxxvii.
5. Ibid., pp. 122–123. Also cited by Davenport, p. 7.
6. *Thirteenth Annual Report of the Secretary of the State Board of Health of the State of Michigan for the Fiscal Year Ending September 30, 1885* (Lansing, MI: W.S. George, 1886), pp. 221–226. Also cited by Davenport, p. 7.
7. Ibid.
8. Ptomaine (Gr. *corpse*) is the common term generically referring to any bacterial toxin. Vaughan initially believed there existed two classes of such toxins: ptomaines and leucomaines, each the products of protein putrefaction. The difference between them was that ptomaines were believed to derive from decomposition of tissue, and leucomaines were a product of tissue metabolism.
9. Ibid., p. 226.
10. Victor C. Vaughan, "Preliminary Note on the Chemistry of Tyrotoxicon," *The Medical News* (April 2, 1887) 50: 369–370.
11. Harold Swithinbank and George Newman, *Bacteriology of Milk* (New York: E.P. Dutton, 1903), p. 208.
12. Catherine J. Whitaker, *The Early Years of the University of Michigan Medical School: The Upjohn Family Experience* (Ann Arbor: Michigan Historical Collections/Bentley Historical Library, 1982), p. 20.
13. Davenport, *op. cit.*, p. 16. Davenport relates that Sewall believed that Behring, later awarded a Nobel Prize in Physiology or Medicine, cheated Sewall of credit for his work.
14. Thomas Neville Bonner, *American Doctors and German Universities* (Lincoln: University of Nebraska Press, 1963), p. 3.
15. Ibid., p. 6.
16. David McCullough, *The Greater Journey: Americans in Paris* (New York: Simon and Schuster, 2011), p. 124.
17. Bonner, *op. cit.*, p. 16.
18. Simon Flexner and James Thomas Flexner, *William Henry Welch and the Heroic Age of American Medicine* (Baltimore: Johns Hopkins University Press, 1993). Other members of the "Big Four" were William Halsted, Sir William Osler and Howard Kelly.
19. Vaughan, *op. cit.*, p. 146.
20. *Fifteenth Annual Report of the Secretary of the State Board of Health of the State of Michigan for the Fiscal Year Ending June 30, 1887* (Lansing, MI: Thorp and Godfrey, 1887), pp. xliv–xlv.

21. Ibid., p. xlv.
22. *Journal of the Senate of the State of Michigan, 1887*, Vol. 2 (Lansing, MI: Thorp and Godfrey, 1887), p. 1659.
23. University of Michigan, Board of Regents, *Proceedings of the Board of Regents* (July 8, 1877), pp. 143-144. The appointment of Vaughan was approved the following day, July 9, 1887.
24. Ibid., pp. 138-139.
25. Ibid., p. 144.
26. *Eighteenth Annual Report of the Secretary of the State Board of Health of the State of Michigan for the Fiscal Year Ending June 30, 1890* (Lansing, MI: Robert Smith, 1892), p. 220.
27. The etiological agent of typhoid fever had been isolated by Karl Eberth in 1880. The genus name was later changed to *Salmonella*, by which it is known today.
28. Victor Vaughan, "A Bacteriological Study of Drinking Water," *Transactions of the Association of American Physicians* 7 (1892), p. 39.
29. Charles Murchison, *A Treatise on the Continued Fevers of Great Britain* (London: Parker, Son and Bourn, 1862). A significant portion of the treatise is used to differentiate typhus from enteric fever (typhoid).
30. Vaughan, *op cit.*, pp. 40-41.
31. *The Physician and Surgeon* 14 (October 1892), p. 470.
32. E. Klein, "An Address on the Etiology of Typhoid Fever," *The British Medical Journal* (October 13, 1894) Vol. 2 (1763), pp. 797-800.
33. Vaughan, *op.cit.*, pp. 42-43. Also cited in Davenport, *op. cit.*, p. 8.

Chapter 4

1. *The Physician and Surgeon* 11 (August 1889), p. 355.
2. Davenport, *op. cit.*, pp. 21-22.
3. Wilfred Shaw (ed.), *The University of Michigan: An Encyclopedic Survey*, Vol. 2 (Ann Arbor: University of Michigan Press, 1951), pp. 888-889; Vaughan, *op. cit.*, pp. 245-246.
4. Vaughan, *op. cit.*, pp. 214-223.
5. R.W. Burmeister et al., "Laboratory-acquired Pneumonic Plague," *Annals of Internal Medicine* (May 1962) 56 (5): 789-800.
6. http://www.ncbi.nlm.nih.gov/pmc/articles/PMC195158/?page=1.
7. Davenport, *op. cit.*, p. 17.
8. Ibid., p. 29.
9. Dale C. Smith, "The Rise and Fall of Typhomalarial Fever," *Journal of the History of Medicine and Allied Sciences* 37 (April 1982) (2): 182-220. http://jhmas.oxfordjournals.org/content/XXXVII/2/182.full.pdf. Also cited in Horace Davenport, *Victor Vaughan: Statesman and Scientist* (Ann Arbor: University of Michigan Historical Center for the Health Sciences, 1996), p. 33.
10. Erik Larson, *The Devil in the White City* (New York: Crown, 2003).
11. William Sedgwick and Allen Hazen, "Typhoid Fever in Chicago," *Engineering News and American Railway Journal* 27 (April 21, 1892), http://www.encyclopedia.chicagohistory.org/pages/10722.html.
12. Ibid., p. 16.
13. Ibid.
14. Vaughan, *op. cit.*, pp. 179-180.
15. Larson, *op. cit.*, pp. 138-139.
16. Davenport (1996), *op. cit.*, p. 33.
17. Arthur Silverstein, "Cellular versus Humoral Immunity: Determinants and Consequences of an Epic 19th Century Battle," *Cellular Immunology* (1979) 48: 208-221.
18. V.C. Vaughan and C.T. McClintock, "The Nature of the Germicidal Constituent of Blood-Serum," *Medical News* 63 (1893): 701-707. Cited by Davenport, *op. cit.*, p. 41.
19. Victor Vaughan, Frederick G. Novy and Charles McClintock, "The Germicidal Properties of Nucleins," *Medical News* 62 (May 20, 1893): 536-538.
20. Davenport, *op. cit.*, p. 42.
21. Ibid.
22. *Nuclein Therapy, Its Rationale, Methods and Results* (Parke, Davis and Company Press, 1900), pp. 19-20.
23. Ibid., pp. 31-33.
24. Matthew Kuchan, Timothy Winship and Marc Masor, "Nucleotides in Infant Nutrition: Effects on Immune Function," *Pediatric Nutrition* 8 (1998): 80-94.
25. Victor C. Vaughan, "The Physiologic Actions and Therapeutic Uses of Yeast Nucleinic Acid, With Special Reference to its Employment in Tuberculosis," *The Medical News* (February 27, 1897) 70 (9): 256-264.
26. George Dock, "The Medical Library of the University of Michigan," *Medical Library and Historical Journal* (July 3, 1905) 3:

165–173; "The Medical Library of the University of Michigan," *The Michigan Alumnus* (March 1907) 13: 240–248.
27. Vaughan, *A Doctor's Memories, op. cit.*, pp. 205–206.
28. Ibid.
29. Dock, *op. cit.*, pp. 166–167.
30. Dock, op. cit., p. 243.
31. http://www.ncbi.nlm.nih.gov/pmc/articles/PMC2909426/.
32. Dock, *op. cit.*, p. 245.

Chapter 5

1. Victor C. Vaughan and Frederick G. Novy, *Ptomaines and Leucomaines, or the Putrefactive and Physiological Alkaloids* (Philadelphia: Lea Brothers, 1888). Cited in Davenport, *op. cit.*, p. 39.
2. Vaughan, *op. cit.*, p. 2.
3. Ibid.
4. Ibid., p. 14.
5. Ibid., pp. 86–87.
6. Ibid., pp. 88–89.
7. Ibid., pp. 89–90.
8. Ibid., p. 90.
9. Ibid., pp. 91–92. Hoffa's work early in his career was in the field of bacteriology. Later he turned to surgery, an area in which he is considered among the founders of modern orthopedics.
10. Thomas Stedman (ed.), *Twentieth Century Practice*, Vol. 13, *Infectious Diseases* (New York: William Wood, 1898), p. 81.
11. Vaughan, *op. cit.*, p. 92.
12. Ibid.
13. Ibid., p. 93. Also cited in Davenport, *op. cit.*, p. 39.
14. Ibid.
15. Charles Chapin, *The Role of Ptomaines in Infectious Disease* (Providence: Snow and Farnham, 1889), p. 22.
16. http://www.unboundmedicine.com/medline/citation/4944138/%5BFrancesco_Selmi_and_the_discovery_of_ptomaine_1870_%5D_.
17. Chapin, *op. cit.*
18. Ibid., pp. 96–97.
19. Willis Alonzo Dewey, *Medical Century: An International Journal of Homeopathic Medicine and Surgery*, Vol. 2 (Chicago, 1894), p. 393.
20. H.M. McClanahan, "Summer Diarrhea of Infants: Its Etiology and Treatment,"
The Journal of the American Medical Association (August 5, 1899) 33 (6): 321.
21. Hobart Amory Hare (ed.), *Progressive Medicine*, Vol. 1 (March 1899): 121.
22. J.A. Larrabee, "Cholera Infantum," *The Medical World* (June 1892) 10 (6): 205–206.
23. Dewey, *op. cit.*
24. Victor C. Vaughan, "Summer Diarrheas of Infancy," *The Medical News* (June 9, 1888) 52 (23): 623–624.
25. Ibid., p. 624. Many of the biochemical tests originally developed by Escherich are still in use over 125 years later. Escherich's work is also the reason for the eponymous name for the colon bacillus.
26. Vaughan's reference is to an outbreak which had taken place in two hotels in Long Branch, Minnesota. For a later description of the outbreak: Hubert Carel, "A Note on Tyrotoxicon Poisoning," *Northwestern Lancet* (April 1, 1904) 24 (7): 122–124.
27. Vaughan, "Summer Diarrheas," *op. cit.*, pp. 624–625.
28. Chapin, *op cit.*, pp. 289–290.
29. Ibid., pp. 292–293. Victor Clarence Vaughan and Frederick G. Novy, *Cellular Toxins, or, the Chemical Factors in the Causation of Disease*, 4th ed. (Philadelphia: Lea Brothers, 1902), p. 483.

Chapter 6

1. Victor C. Vaughan, *A Doctor's Memories* (Indianapolis: Bobbs-Merrill, 1926), pp. 327–328.
2. Merrite Ireland, "A Fighter for the Cause of Health," *The Journal of Laboratory and Clinical Medicine* 15 (9) (June 1930): 878–884.
3. Vaughan, *op. cit.*, p. 328.
4. Ibid., p. 341.
5. Ibid., pp. 357–358.
6. Vincent Cirillo, "Fever and Reform: The Typhoid Epidemic in the Spanish-American War," *Journal of the History of Medicine and Allied Sciences* 55 (4) (October 2000): 363–397.
7. Bobby Wintermute, *Public Health and the U.S. Military: A History of the Army Medical Department, 1818–1917* (New York: Routledge, 2011), p. 95.
8. Ibid., p. 96.
9. Cirillo, *op. cit.*, p. 370.

10. Walter Reed, Victor Clarence Vaughan and Edward Oram Shakespeare, *Report on the Origin and Spread of Typhoid Fever in United States Military Camps During the Spanish War of 1898* (Washington, D.C.: Government Printing Office, 1904), p. xv; Also cited in Cirillo, *op. cit.*, p. 370. Henry Clark Corbin, the adjutant general, was cited for his "gallant and meritorious service" as an officer during the Civil War. He was with President James Garfield when the latter was shot by an assassin in 1881, and was by the bedside when the president later died.
11. Vaughan, *op. cit.*, pp. 370–392.
12. Ibid., pp. 370–371.
13. Reed, *op. cit.*
14. Vaughan, *op. cit.*, pp. 372–373.
15. Ibid., pp. 371–372.
16. Cirillo, *op. cit.*, p. 367.
17. Bobby Wintermute, *Public Health and the U.S. Military: A History of the Army Medical Department, 1818–1917* (New York: Routledge, 2011), pp. 83–85.
18. Harry E. Webber, *Twelve Months with the Eighth Massachusetts Infantry in the Service of the United States* (Newcomb and Gauss, 1908), p. 87; cited by Wintermute, *op. cit.*, p. 86.
19. Wintermute, *op. cit.*, p. 86.
20. Reed, *op. cit.*; Wintermute, *op. cit.*, p. 104.
21. Horace Davenport, *Not Just Any Medical School* (Ann Arbor: University of Michigan Press, 1999), p. 44; Vaughan, *op. cit.*, p. 380.
22. Wintermute, *op. cit.*, p. 104.
23. Ibid., p. 105.
24. Vaughan, *op. cit.*, p. 389.
25. William Budd, "Typhoid or Intestinal Fever: The Pythogenic Theory," *British Medical Journal* (November 16, 1861) 2 (46): 523–525. Budd was among the earliest (ca. 1839) to propose typhoid fever was spread through contaminated water.
26. Reed, *op. cit.*, p. 662.
27. Ibid.
28. Ibid., p. 663.
29. Ibid.
30. Ibid., p. 293.
31. Ibid., p. 666.
32. Ibid., p. 665.
33. Alice Hamilton, "The Fly as a Carrier of Typhoid: An Inquiry into the Part Played by the Common House Fly in the Recent Epidemic of Typhoid Fever in Chicago," *The Journal of the American Medical Association* (February 28, 1903) 15 (9): 576–583.
34. Reed, *op. cit.*, p. 664.
35. Ibid., p. 665.
36. Vaughan, *op. cit.*, p. 390.
37. Reed, *op. cit.*, p. 674.
38. http://history.amedd.army.mil/booksdocs/wwii/PM4/CH22.Typhoid.htm.
39. *Thucydides' Peloponnesian War Book II*, Chapter 7, 2.47–2.55.
40. Manolis Papagrigorakis et al., "Typhoid Epidemic in Ancient Athens," pp. 161–173. In D. Raoult and M. Drancourt (eds.), *Paleomicrobiology: Past Human Infections* (Berlin: Springer-Verlag, 2008).
41. Burke Cunha, "Osler on Typhoid Fever: Differentiating Typhoid from Typhus and Malaria," *Infectious Disease Clinics of North America* (2004) 18: 111–125.
42. Ibid., p. 112.
43. Ibid.
44. Frederick Parker Gay, *Typhoid Fever Considered as a Problem of Scientific Medicine* (New York: Macmillan, 1918), p. 5.
45. W. Jenner, *On Identity or Non-identity of Typhoid and Typhus Fevers* (London: John Churchill, 1850), p. 95.
46. Robert Dorfman et al. (eds.), *Models for Managing Regional Water Quality* (Cambridge: Harvard University Press, 1972), p. 3.
47. William Budd, "Typhoid Fever: Its Nature, Mode of Spreading, and Prevention," *The American Journal of Public Health* (August 1918) 8 (8): 610–612 (originally published in 1873).
48. Gay, *op. cit.*, p. 8.
49. J. Stachura and K. Galazka, "History and Current Status of Polish Gastroenterological Pathology," *Journal of Physiology and Pharmacology* (2003) 54 (3): 183–192. Browicz later identified the Kupffer cells in the liver.
50. Gay, *op. cit.*, p. 9; G. Gaffky, "Zur aetiologie des abdominaltyphus" ("On the etiology of abdominal typhus"), *Mittheilungen aus den Kaiserlichen Gesundheitsamte*, Vol. 2 (Berlin, 1884), pp. 372–420.
51. Fernand Widal, "On the Sero-Diagnosis of Typhoid Fever," *The Lancet* (November 14, 1896) 2 (3820): 1371–1372.
52. Albert Grünbaum, "Preliminary Note on the Use of the Agglutinative Action of Human Serum for the Diagnosis of En-

teric Fever," *The Lancet* (September 19, 1896) 2 (3812): 806–807.
53. Widal, *op. cit.*, p. 1372.
54. Ibid.

Chapter 7

1. Davenport, *op. cit.*, p. 25.
2. Ibid., p. 51.
3. http://coursesa.matrix.msu.edu/~hst425/readings/Berliner.htm.
4. Ibid.
5. Arthur Dean Bevan, "Cooperation in Medical Education and Medical Service," *The Journal of the American Medical Association* (April 14, 1928) 90 (15): 1173–1177.
6. "Council on Medical Education," *The Journal of the American Medical Association* (January 14, 1905) 44 (2): 147.
7. Bevan, *op. cit.*
8. Vaughan, *op. cit.*, p. 440; cited in Davenport, *op. cit.*, p. 52.
9. Abraham Flexner, *Medical Education in the United States and Canada* (New York: Carnegie Foundation, 1910), pp. x–xi.
10. Vaughan, *op. cit.*, pp. 442–443.
11. As the author was writing this he came upon an ad for "prostate health" which included the disclaimer: "Results may vary. If you are pregnant, nursing, taking other medications, or have a serious medical condition, we suggest consulting with a physician before using a prostate health supplement."
12. "A Typical Anti-vaccinationist," *The Journal of the American Medical Association* (July 14, 1894) 23(2), p. 84.
13. Davenport, *op. cit.*, p. 44.
14. Ibid.
15. *Contributions to Medical Research Dedicated to Victor Clarence Vaughan* (Ann Arbor, MI: George Wahr, 1903).
16. "The Twenty-fifth Anniversary of Dr. Victor C. Vaughan's Graduation," *The Journal of the American Medical Association* (July 4, 1903) 41 (1): 48.
17. Davenport, *op. cit.*, p. 91.
18. H.C. Gram, "Über die isolierte Färbung der Schizomyceten in Schnitt- und Trockenpräparaten" ("Differential staining of Schizomycetes in Sections and in Smear Preparations"), *Fortschritte der Medizin* 2 (1884): 185–189.
19. Ibid.
20. C. Flugge, *The Micro-organisms with Special Reference to the Etiology of the Infective Diseases*, W. Watson Cheyne, trans. (London: New Sydenham Society, 1890). Also cited by Hubert A. Lechevalie and Morris Solotorovsky, *Three Centuries of Microbiology* (New York: Dover, 1974), p. 137.
21. Victor C. Vaughan and Thomas B. Cooley, "The Bacterial Toxins," *The Journal of the American Medical Association* (February 23, 1901) 36 (8): 479–482. Quoted by Davenport, *op. cit.*, p. 92.
22. Victor C. Vaughan, *A Contribution to the Chemistry of the Bacterial Cell and a Study of the Effects of Some of the Split Products on Animals*, presentation at the Annual Meeting of the Massachusetts Medical Society, June 12, 1906, p. 22; reprinted in *Boston Medical and Surgical Journal* (August 30, 1906) 155 (9): 215–222.
23. Ibid., p. 23.
24. Ibid., p. 31.
25. R. Pfeiffer, "Untersuchungen über das Choleragift" ("Studies of the Cholera Poison"), *Zeitschrift für Hygiene* (*Journal of Hygiene*) (1892) 11: 393–411.
26. Helmut Brade (ed.), *Endotoxin in Health and Disease* (New York: Marcel Decker, 1999), p. 8.
27. Hans Zinsser, *Infection and Resistance*, 2nd ed. (New York: Macmillan, 1922), pp. 37–38.
28. Vaughan, *Boston Medical and Surgical Journal, op. cit.*, p. 220.
29. Brade, *op. cit.* pp. 9–10; A. Boivin and L. Mesrobeanu, "Recherches sur les antigenes somatiques et sur les endotoxins les bacteries. I. Considerations generales et expose des technique utilisees," *Revue d'Immunologie* (1935) 1: 553–569.
30. Victor C. Vaughan and Frederick G. Novy. *Cellular Toxins; or, the Chemical Factors in the Causation of Disease*, 4th ed. (Philadelphia: Lea Brothers, 1902).
31. Ibid., p. 23.
32. Ibid., p. 20.
33. Ibid., p. 22.
34. Ibid.
35. Victor C. Vaughan and Frederick G. Novy, *Ptomaines, Leucomaines, Toxins and Antitoxins or the Chemical Factors in the Causation of Disease*, 3rd ed. (Philadelphia: Lea Brothers, 1896), p. 135.
36. Vaughan, *Cellular Toxins, op. cit.*, pp. 22–23.

37. Ibid., pp. 23-24.
38. Jeongmin Song et al., "Structure and Function of the *Salmonella typhi* chimaeric A2B5 typhoid toxin," http://www.nature.com/nature/journal/vaop/ncurrent/full/nature12377.html.
39. Vaughan, *op. cit.*, p. 102.
40. Ibid.
41. Franklin Warren White, "Experiments Upon the Germicidal Properties of Blood Serum," *The Boston Medical and Surgical Journal* (February 23, 1899) 40 (8): 177-183.
42. Vaughan, *op. cit.*, pp. 103-104.
43. Ibid., pp. 107-108.
44. Ibid., p. 109.
45. Ibid., p. 165.
46. Ibid., pp. 165-166.
47. Vaughan, *Ptomaines and Leucomaines, op. cit.*, pp. 86-87.
48. Vaughan, *Cellular Toxins, op. cit.*, p. 166.
49. Ibid., p. 168.
50. Ibid., p. 169.
51. Davenport, *op. cit.* p. 41.; Vaughan, *op. cit.*, pp. 179, 183.
52. Vaughan, *op. cit.*, p. 181.
53. Ibid., pp. 485-486. In the present day the latter is often referred to as the Thanksgiving meal.
54. Ibid., p. 485.
55. Ibid., p. 486. Myxedema refers to the improper deposition of connective tissue derivatives under the surface of skin. While the cause may vary, it is often associated with hypothyroidism.
56. Ibid., pp. 486-487.
57. "Alleged Rabies (Hydrophobia) in Michigan in 1900," *Twenty-ninth Annual Report of the Secretary of the State Board of Health of the State of Michigan for the Fiscal Year Ending June 30, 1901* (Lansing, MI: Wynkoop Hallenbeck Crawford, 1902), Part 2, p. 253.
58. *The Journal of the Michigan State Medical Society* (May 1903) 11(5): 222.
59. Davenport, *op. cit.*, p. 41.
60. Victor C. Vaughan, "How Bacteria Cause Disease," in *Transactions of the Clinical Society of the University of Michigan*. Vol. 5 (February 11, 1914), p. 59.
61. Victor C. Vaughan, "The Protein Poison in Health and Disease," *Journal of the Medical Society of New Jersey* (November 1917) 14 (11): 417-424.

62. Victor Vaughan, "The Michigan Method of Water Analysis," *Public Health, Michigan* (January-March 1912) 7 (1): 9-24.
63. Ibid., p. 11.
64. Ibid., p. 16.
65. Vaughan, "How Bacteria Cause Disease," *op. cit.*, p. 60.
66. Vaughan, "The Protein Poison in Health and Disease," *op. cit.* Friedberger, a German veterinarian, believed ptomaines and such poisons were most likely the result of bacterial metabolism rather than the products of a protein digestion. Franz Friedberger et al., *Friedberger and Frohner's Veterinary Pathology* Vol. 2 (authorized translation), 6th ed. (Chicago: W.T. Keener, 1908).
67. Victor Vaughan, "The Protein Poison," *Proceedings of the American Philosophical Society* 51 (Lancaster, PA: New Era Printing, 1912), p. 209.
68. Ibid., pp. 209-210.
69. Ibid., pp. 211-212.
70. Ibid., pp. 212-213.
71. Victor C. Vaughan and Sybil May Wheeler, "The Effects of Egg-white and its Split Products on Animals: A Study of Susceptibility and Immunity," *The Journal of Infectious Diseases* (June 1907) 4: 476-508.
72. Ibid.
73. *The Journal of the American Medical Association* (August 3, 1912) 59 (5): 372. Though his proposal was premature, Friedberger was not incorrect in suggesting a role for antibody in the anaphylactic reaction. Anaphylaxis had first been described by Charles Richet and Paul Portier in 1901; Richet coined the term to represent a "lack of protection." It was only in the 1920s that a specific antibody, then referred to as a reagin but now known as IgE, was observed in the serum of sensitized animals. The antibody is not an enzyme, but functions in the release of vasoactive mediators from certain cells.
74. V.C. Vaughan, *The Journal of the American Medical Association* (August 17, 1912) 59 (7): 563.
75. Victor C. Vaughan, "How Bacteria Cause Disease," *op. cit.*, pp. 63-65.
76. Ibid., p. 67.
77. Victor C. Vaughan, "Protein Fever," in *Poisonous Proteins: The Herter Lectures for 1916* (St. Louis: C.V. Mosby, 1917), pp. 75-79.

Chapter 8

1. Joel D. Howell, *Medical Lives and Scientific Medicine at Michigan, 1891–1969* (Ann Arbor: University of Michigan Press, 1993), p. 54.
2. Ibid., pp. 56–57.
3. Ibid., p. 60.
4. Ibid., p. 61.
5. Ibid., p. 63. Much the same argument continues a century later when the university justifies the proffering of significant salaries when recruiting for administrative positions.
6. Ibid., p. 65.
7. Randall Hansen and Desmond King, *Sterilized by the State: Eugenics, Race and the Population Scare in Twentieth-century North America* (New York: Cambridge University Press, 2013), p. 85.
8. Victor C. Vaughan, *A Doctor's Memories* (Indianapolis: Bobbs-Merrill, 1926), p. 422.
9. http://www7.nationalacademies.org/ocga/other/Act_to_incorporate.asp.
10. Rexmond Cochrane, *The National Academy of Sciences: The First 100 Years, 1863–1963* (Washington, D.C.: National Academy of Sciences, 1978), p. 208.
11. Ibid., pp. 208–209.
12. "The National Research Council," *Science* (August 25, 1916) 44 (1130): 264–266.
13. Ibid.
14. Ibid.
15. Cochrane, *op. cit.*, p. 215.
16. *First Annual Report of the Council of National Defense* (Washington, D.C., 1917), p. 48.
17. Victor C. Vaughan, *op. cit.*, p. 403.

Chapter 9

1. *Official Bulletin, United States Committee on Public Information* (May 11, 1917) 1 (2): 6. (Hereafter referred to as *Official Bulletin*.)
2. Victor Vaughan, *A Doctor's Memories* (Indianapolis: Bobbs-Merrill, 1926), pp. 410–411.
3. Ibid.
4. *Official Bulletin* (July 30, 1917) 1 (68): 8.
5. *Official Bulletin* (July 28, 1917) 1 (67): 3.
6. Carol R. Byerly, *Fever of War: The Influenza Epidemic in the U.S. Army during World War I* (New York: New York University Press, 2005), p. 54.
7. Vaughan, *op. cit.*, p. 424.
8. "Gorgas Reports Troops at Camps Crowded, Ill-Clad," *The New York Times* (December 19, 1917): 1.
9. Byerly, *op. cit.*, pp. 59–61.
10. George Soper, "The Influenza Pneumonia Pandemic in the American Army Camps During September and October, 1918," *Science* (November 8, 1918) 48 (1245): 451. Also cited by Byerly, *op. cit.*, p. 88. Soper in his civilian life had been a major figure in uncovering the role of Mary Mallon, "Typhoid Mary," in 1907.
11. W. Beveridge, "The Chronicle of Influenza Epidemics," *History and Philosophy of the Life Sciences* (1991) 13: 223–235; John Townsend, *op. cit.*; Victor Clarence Vaughan, *Epidemiology and Public Health: A Text and Reference Book for Physicians*, Vol. 1 (St. Louis, MO: C.V. Mosby, 1922), pp. 297–313 (prior to the 1918 outbreak).
12. K. David Patterson, *Pandemic Influenza: 1700–1900* (Totowa, NJ: Rowman and Littlefield, 1986).
13. John Barry, *The Great Influenza: The Story of the Greatest Pandemic in History* (New York: Penguin Group, 2005); Gina Kolata, *Flu: The Story of the Great Influenza Pandemic* (New York: Touchstone, 1999).
14. Edwin Kilbourne, *Influenza* (New York: Plenum Medical Book Company, 1987), p. 4.
15. Vaughan, *op. cit.*, p. 312.
16. Arthur Silverstein, *A History of Immunology*, 2nd ed. (New York: Elsevier, 2009).
17. J.K. Taubenberger and D.M. Morens, "Pandemic influenza—including a risk assessment of H5N1," *Revue Scientifique et Technique* (April 2009) 28 (1): 187–202.
18. Beveridge, *op. cit.*, pp. 223–224.
19. William Gordon, *Titus Livius's Roman History Translated into English*, Vol. 2 (Glasgow: William Smith, 1783), p. 78; Cited in Arthur Hopkirk, *Influenza: Its History, Nature, Cause and Treatment* (New York: Charles Scribner's Sons, 1914), p. 16.
20. John Townsend, "History of Influenza Epidemics," *Annals of Medical History* (1933) 6: 533–547. The Siege of Syracuse was the first of four such (unsuccessful) sieges by Carthage.

21. Arthur Hopkirk, *op. cit.*, p. 19.
22. Townsend, *op. cit.*, p. 536; Beveridge, *op. cit.*, p. 225.
23. Thomas Stedman (ed.), *Twentieth Century Practice of Modern Medical Science: Infectious Diseases*, Vol. 15 (New York: William Wood, 1898), p. 5. A caveat must be included here. Stedman does not conflate the 876 outbreak with movement of Charlemagne's armies. This differs from several other sources, including Beveridge and Taubenberger, who indicate the outbreak, whatever it might have been, followed the armies of Charlemagne. Charlemagne, king of the Franks and first Holy Roman emperor, died ca. 814, long before the outbreak. To clarify names and dates, this author chose to contact the historians directly. William Beveridge died in 2006. Dr. Taubenberger was kind enough to answer my inquiry with the following reply: "'Our' English language Charlemagne [Karolmann] lived from 742–814, but the Karolman who wrote about the flu epidemic in 876–877 was a different Karolman who lived 828 or 830 to 880 and was sometimes called in French Charles III or Charles le Gros (Charles the Fat, not Charles the Great). The most famous report of that epidemic is the original in the Annales Fuldenses, as popularized by Schnurrer." Reference: Schnurrer F. Chronik der Seuchen, in *Verbindung mit den gleichzeitigen Vorgängen in der physischen Welt und in der Geschichte der Menschen* (Tübingen: Osiander, 1823), p. 1825.
24. Ibid.
25. Taubenberger, *op. cit.*, p. 189.
26. August Hirsch, *Handbook of Geographical and Historical Pathology: Acute Infectious Diseases*, Vol. 1, Charles Creighton, trans. (London: New Sydenham Society, 1883), p. 7.
27. Ibid., p. 18.
28. W.T. Vaughan, "Influenza, an Epidemic Study," *American Journal of Hygiene* (1921), Monograph Series No. 1, Baltimore, MD; Beveridge, *op. cit.*, p. 225.
29. Gottlieb Gluge, "*Die Influenza oder Grippe, nach den Quellen Historisch-Pathologisch Dargestellt*" (Minden: Verlag von Ferdinand Efsmann, 1837).
30. Ibid., pp. 26–35.
31. Taubenberger, *op. cit.*, p. 189.
32. Hopkirk, *op. cit.*, pp. 21–22.
33. Richard Sisley, *Epidemic Influenza: Notes on its Origin and Method of Spread* (London: Longmans, Green, 1891), p. 4.
34. Gluge, *op. cit.*, p. 16.
35. Etienne Pasquier, *Recherches de la France (Research in French History)* (Paris, 1596). Townsend also ascribes the year as 1413. Townsend, *op. cit.*, p. 536.
36. Hopkirk, *op. cit.*, p. 23.
37. Ibid., p. 24; Stedman, *op. cit.*, p. 9.
38. Hopkirk, *op. cit.*, p. 24.
39. Stedman, *op. cit.*, p. 9.
40. Hopkirk, *op. cit.*, p. 24.
41. Stedman, *op. cit.*, p. 9.
42. Ibid., p. 10.
43. Ibid., p. 10; Hopkirk, *op. cit.*, p. 25.
44. Alexander Lambert, "Influenza," *New York State Journal of Medicine* (July 1919) 19 (7): 260; Theophilus Thompson (ed.), *Annals of Influenza or Epidemic Catarrhal Fever in Great Britain from 1510 to 1837* (London: Sydenham Society, 1852), p. 3.
45. David Morens, Michael North, and Jeffery Taubenberger. "Eyewitness Accounts of the 1510 Influenza Pandemic in Europe," *The Lancet* (December 4, 2010) 376 (9756): 1894–1895.
46. Stedman, *op. cit.*, p. 12.
47. Ibid.
48. Ibid., p. 13.
49. Thompson, *op. cit.*, pp. 6–7.
50. Taubenberger, *op. cit.*, p. 5.
51. Thompson, *op. cit.*, p. 8.
52. Vaughan, *op. cit.*, p. 301.
53. Stedman, *op. cit.*, p. 14.
54. Ibid.
55. Ibid., p. 16.
56. Gluge, *op. cit.*, p. 60.
57. Stedman, *op. cit.*, p. 16.
58. Gluge, *op. cit.*, p. 60.
59. David Morens, Jeffery Taubenberger, Gregory Folkers and Anthony Fauci, "Pandemic Influenza's 500th Anniversary," *Clinical Infectious Diseases* (December 15, 2010) 51: 1442–1444.
60. Thompson, *op. cit.*, pp. 8–10.
61. Hopkirk, *op. cit.*, p. 31.
62. Hermanni Boerhaave, *Methodus Studii Medici, Emaculata, & Accessionibus Locpletata* (Amstelaedaml: Jacobi a Wetstein, 1751), p. 862.
63. Hopkirk, *op. cit.*, p. 32.
64. Stedman, *op. cit.*, p. 16.
65. Robley Dunglison, *The Practice of Medicine: A Treatise on Special Pathology and Therapeutics*, 3rd ed. (Philadelphia: Lea

and Blanchard, 1848), p. 278; Gluge, *op. cit.*, pp. 18–19.

66. Hopkirk, *op. cit.*, p. 32.

67. *Journal of the American Medical Association* (February 23, 1920) 74 (9): 606; Guy Hinsdale, "Epidemics of Influenza in 1647, 1789–90 and 1807 as Recorded by Noah Webster, Benjamin Rush and Daniel Drake," in *Contributions to Medical and Biological Research Dedicated to Sir William Osler*, Vol. 2 (New York: Paul B. Hoeber, 1919), p. 721. Webster's description is recounted above. Benjamin Rush was a Philadelphia physician and among the signers of the Declaration of Independence. Rush's account of the influenza epidemic of 1789 and 1790 includes symptoms such as a sore throat, fever, headache and severe joint pains, often followed by pneumonia and a high mortality rate, a description which could easily apply to the 1918 pandemic. Drake, an army surgeon stationed in the Ohio Valley, described an outbreak of influenza which took place during the 1810s.

68. Noah Webster, "On the Origin of Bilious Plague," in *The Medical Repository and Review of American Publications on Medicine, Surgery and the Auxiliary Branches of Science*, Samuel Mitchell and Edward Miller (eds.), Vol. 1 (New York: T. and J. Swords, 1804), p. 322. (See also n. 59.)

69. Noah Webster, *A Brief History of Epidemic and Pestilential Diseases*, Vol. 1 (Hartford, CT: Hudson and Goodwin, 1799), p. 193.

70. http://english.byu.edu/facultysyllabi/KLawrence/WEBSTER.briefhistory.pdf.

71. Joshua Kendall, *The Forgotten Founding Father: Noah Webster's Obsession and the Creation of an American Culture*. (New York: Penguin Group, 2010), p. 212.

72. Patterson, *op. cit.*; Youri Ghendon, "Introduction to Pandemic Influenza Through History," *European Journal of Epidemiology* (1994) 10: 451–453.

73. Ibid., p. 13.

74. Theophilus Thompson, *op. cit.*, pp. 27–28; Kilbourne, *op. cit.* p. 6.

75. Patterson, *op. cit.*, pp. 13–14.

76. Beveridge, *op. cit.*, p. 226.

77. Ibid.

78. Patterson, *op. cit.*, pp. 21–22.

79. Ibid., pp. 26–27.

80. John Huxham, *An Essay on Fevers*, 3rd ed. (London: J. Hinton, 1757), cited by Margaret Delacy, "The Conceptualization of Influenza in Eighteenth Century Britain: Specificity and Contagion," pp. 74–117.

81. Delacy, *op. cit.*, p. 110.

82. Patterson, *op. cit.*, pp. 37–38.

83. Ibid., pp. 39–41.

84. Charles Creighton, *A History of Epidemics in Britain* (Cambridge: The University Press, 1894), p. 390.

85. Patterson, *op. cit.*, pp. 52–53. A detailed description of the spread of the pandemic can be found in Patterson.

86. Ibid., pp. 69–71.

87. Ibid., p. 82.

88. R. Pfeiffer, "*Vorlaufige Mittheil über die Erreger der Influenza*" ("Preliminary Announcement about the Pathogen of Influenza") *Deutsche Medicinische Wochenschrift* (January 1892) 18 (2): 28.

89. "The 1918–1919 Influenza Pandemic as Covered in *The Journal of Immunology* from 1919–1921," *American Association of Immunologists Newsletter* (July-August 2012): 12–13.

90. C. Roos, "Notes on the Bacteriology, and on the Selective Action of B. influenzae Pfeiffer," *The Journal of Immunology* (July 1, 1919) 4 (4): 189–201.

91. Eugenia Valentine and Georgia M. Cooper, "On the Existence of a Multiplicity of Races of B. influenzae as Determined by Agglutination and Agglutinin Absorption," *The Journal of Immunology* (September 1, 1919) 4 (5): 359–379.

92. Kolata, *op. cit.*, p. 67.

93. Richard E. Shope, "Swine Influenza. III. Filtration Experiments and Etiology," *Journal of Experimental Medicine* (July 31, 1931) 54 (3): 373–385.

94. Wilson Smith, C.H. Andrewes, and P.P. Laidlaw, "A Virus Obtained from Influenza Patients," *The Lancet* (July 8, 1933) 222 (5732): 66–68.

95. Ibid., p. 68.

96. Victor Clarence Vaughan, Henry Frieze Vaughan and George Truman Palmer, *Epidemiology and Public Health*, Vol. 1 (St. Louis: C.V. Mosby, 1922), p. 313.

97. http://www.doctorsreview.com/history/nov05-history/; http://www.mocavo.com/Albert-Gitchell-1890-1968-Social-Security-Death-Index/16667150139699329692.

98. *Corrections Response to Pandemic Flu*, www.asca.net/.../_ASCA_Pandemic_

Planning_Guide_2010_1_.pdf?, p. 2. Camp Funston was named in honor of Major General Frederick Funston, a Medal of Honor recipient for action during the war in the Philippines. Funston had died suddenly from a heart attack in January 1917 at age fifty-one.

99. Byerly, *op. cit.*, p. 73.
100. Ibid.
101. "The Prevailing Epidemic," *The Lancet* (July 13, 1918) 192 (4950): 51.
102. Captain T.R. Little et al. "The Absence of the Bacillus influenzae in the Exudate from the Upper Air-Passages in the Present Epidemic," *The Lancet* (July 13, 1918) 192 (4950): 34.
103. Byerly, *op. cit.*, pp. 74–75.
104. John Barry, *op. cit.*, p. 187.
105. Ibid.
106. Simon Flexner and James Thomas Flexner, *William Henry Welch and the Heroic Age of American Medicine* (Baltimore: Johns Hopkins University Press, 1941), p. 372.
107. Vaughan, *Epidemiology, op. cit.*, p. 8.
108. Vaughan, *Memories, op. cit.*, pp. 383–384.
109. Flexner, *op. cit.*, p. 376.
110. Byerly, *op. cit.*, pp. 78–79.
111. Barry, *op. cit.*, pp. 190–191.
112. *Review of War Surgery and Medicine*, Vol. 2 (1) (Washington, D.C.: Office of the Surgeon General, 1919), p. 84.
113. Wolbach, http://collections.nlm.nih.gov/ocr/nlm:nlmuid-101467619-bk.
114. Soper, *op. cit.*
115. Barry, *op. cit.*, p. 290.
116. V.C. Vaughan, "Notes on Influenza," *The Journal of Laboratory and Clinical Medicine* (December 18, 1918) 4 (3): 145–148.
117. Soper, *op. cit.*, p. 454.
118. Byerly, http://www.ncbi.nlm.nih.gov/pmc/articles/PMC2862337/.
119. John Barry, Cecile Viboud, and Lone Simonsen, "Cross-protection Between Successive Waves of the 1918–1919 Influenza Pandemic: Epidemiologic Evidence from U.S. Army Camps and From Britain," *Journal of Infectious Diseases* (November 15, 2008) 198: 1427–1434.
120. Robert Mearns Yerkes (ed.), *The New World of Science: Its Development During the War* (New York: Century, 1920), p. 332.
121. Ibid., p. 354.

Chapter 10

1. *Proceedings of the University of Michigan Board of Regents* (March 25, 1921), p. 149–150.
2. *Report of the National Academy of Sciences Fiscal Year 1924–1925* (Washington, D.C.: Government Printing Office, 1926), pp. 221–222.
3. Victor Clarence Vaughan, *Epidemiology and Public Health*, Vol. 1 (St. Louis, MO: C.V. Mosby, 1922), pp. 18–19.
4. Ibid., p. 23.
5. Ibid., p. 44.
6. Ibid., pp. 49–50. In 1922 the concept of tissue specific histocompatibility antigens was unknown. It is likely the results of the guinea pig experiments to which Vaughan alluded were due to anaphylactic shock induced by foreign proteins in animals not inbred. The results of the prodigiosus experiment may have resulted from endotoxin released from the Gram-negative cell. Or just as likely, given that the greater the immunity the lesser the lethal dose, Vaughan may have observed anaphylaxis.
7. Ibid., pp. 58–59. The differences in pathogenic potential among these examples would not be understood for decades after Vaughan's death. For example, corynebacteria associated with diphtheria only become pathogenic as a result of lysogenic conversion, the integration of a specific bacteriophage which encodes the toxin gene.
8. Ibid., p. 69.
9. Ibid., p. 76. The wet curtains would have functioned in collecting the pollen grains, not in the removal of proteins.
10. Ibid., pp. 76–77.
11. Jeffrey Jentzen, *Death Investigation in America* (Cambridge, MA: Harvard University Press, 2010), pp. 34–36.
12. *Proceedings of the American Chemical Society* (May 1927), p. 29.
13. E.C.C. Baly, "Photosynthesis," *Nature* (March 16, 1922) 109 (2733): 345.
14. Victor C. Vaughan, "A Chemical Concept of the Origin and Development of Life," *Chemical Reviews* 4 (2): 179–180.
15. Ibid., p. 185.

Bibliography

Books

Barry, John. *The Great Influenza: The Story of the Greatest Pandemic in History*. New York: Penguin Group, 2005.

Bonner, Thomas Neville. *American Doctors and German Universities*. Lincoln: University of Nebraska Press, 1963.

Brade, Helmut (ed.). *Endotoxin in Health and Disease*. New York: Marcel Decker, 1999.

Byerly, Carol R. *Fever of War: The Influenza Epidemic in the U.S. Army during World War I*. New York, NY: New York University Press, 2005.

Chapin, Charles. *The Role of Ptomaines in Infectious Disease*. Providence, RI: Snow and Farnham Printers, 1889.

Cochrane, Rexmond. *The National Academy of Sciences: The First 100 Years, 1863–1963*. Washington, D.C.: National Academy of Sciences, 1978.

Contributions to Medical Research Dedicated to Victor Clarence Vaughan. Ann Arbor, MI: George Wahr, 1903.

Creighton, Charles. *A History of Epidemics in Britain*. Cambridge: The University Press, 1894.

Cushing, Harvey. *The Life of Sir William Osler*. Hamburg, Germany: Severus Verlag, 2010.

Davenport, Horace. *Not Just Any Medical School: The Science, Practice and Teaching of Medicine at the University of Michigan, 1850–1941*. Ann Arbor: University of Michigan Press, 1999.

Davenport, Horace. *Victor Vaughan: Statesman and Scientist*. Ann Arbor: University of Michigan Historical Center for the Health Sciences, 1996.

Dewey, Willis Alonzo. *Medical Century: An International Journal of Homeopathic Medicine and Surgery*. Vol. 2. Chicago, 1894.

Dorfman, Robert, et al. (eds.). *Models for Managing Regional Water Quality*. Cambridge: Harvard University Press, 1972.

Douglas, Silas Hamilton, Samuel Townsend Douglas, Ashley Pond, University of Michigan, Board of Regents. *The Regents of the University of Michigan vs. Preston B. Rose, Silas H. Douglas, Appellant*. State of Michigan, Supreme Court. In Chancery. 1881.

Dunglison, Robley. *The Practice of Medicine: A Treatise on Special Pathology and Therapeutics*. 3rd ed. Philadelphia: Lea and Blanchard, 1848.

Eighteenth Annual Report of the Secretary of the State Board of Health of the State of Michigan for the Fiscal Year Ending June 30, 1890. Lansing, MI: Thorp and Godfrey, 1890.

Fifteenth Annual Report of the Secretary of the State Board of Health of the State of Michigan for the Fiscal Year Ending June 30, 1887. Lansing, MI: Thorp and Godfrey, 1887.

Flexner, Abraham. *Medical Education in the United States and Canada*. New York: Carnegie Foundation, 1910.

Flexner, Simon, and James Thomas Flexner. *William Henry Welch and the*

Heroic Age of American Medicine. Baltimore, MD: Johns Hopkins University Press, 1993.

Flugge, C. *The Micro-organisms with Special Reference to the Etiology of the Infective Diseases*. W. Watson Cheyne, trans. London: New Sydenham Society, 1890.

Friedberger, Franz, et al. *Friedberger and Frohner's Veterinary Pathology*, Vol. 2 (authorized translation). 6th ed. Chicago: W.T. Keener, 1908.

Gay, Frederick Parker. *Typhoid Fever Considered as a Problem of Scientific Medicine*. New York: Macmillan, 1918.

Gordon, William. *Titus Livius's Roman History Translated into English*, Vol. 2. Glasgow: William Smith, 1783.

Hansen, Randall, and Desmond King. *Sterilized by the State: Eugenics, Race and the Population Scare in Twentieth-Century North America*. New York, NY: Cambridge University Press, 2013.

Hirsch, August. *Handbook of Geographical and Historical Pathology: Acute Infectious Diseases*, Vol. 1. Charles Creighton, trans. London: New Sydenham Society, 1883.

Hopkirk, Arthur. *Influenza: Its History, Nature, Cause and Treatment*. New York: Charles Scribner's Sons, 1914.

Howell, Joel D. *Medical Lives and Scientific Medicine at Michigan, 1891–1969*. Ann Arbor: University of Michigan Press, 1993.

Hubbard, William N., and Nicholas H. Steneck. *The Origins of Michigan's Leadership in the Health Sciences*. Ann Arbor, MI: Historical Center for the Health Sciences, 1995.

Huxham, John. *An Essay on Fevers*, 3rd ed. London: J. Hinton, 1757.

Imber, Gerald. *Genius on the Edge: The Bizarre Double Life of Dr. William Stewart Halsted*. New York, NY: Kaplan, 2010.

Jenner, W. *On Identity or Non-Identity of Typhoid and Typhus Fevers*. London: John Churchill, 1850.

Jentzen, Jeffrey. *Death Investigation in America*. Cambridge, MA: Harvard University Press, 2010.

Kendall, Joshua. *The Forgotten Founding Father: Noah Webster's Obsession and the Creation of an American Culture*. New York: Penguin Group, 2010.

Kilbourne, Edwin. *Influenza*. New York: Plenum Medical Book Company, 1987.

Kolata, Gina. *Flu: The Story of the Great Influenza Pandemic*. New York: Touchstone, 1999.

Larson, Erik. *The Devil in the White City*. New York: Crown, 2003.

Lechevalier, Hubert A., and Morris Solotorovsky. *Three Centuries of Microbiology*. New York: Dover, 1974.

McCullough, David. *The Greater Journey, Americans in Paris*. New York, NY: Simon and Schuster, 2011.

Mitchell, Samuel, and Edward Miller (eds.). *The Medical Repository and Review of American Publications on Medicine, Surgery and the Auxiliary Branches of Science*, Vol. 1. New York: T. & J. Swords, 1804.

Murchison, Charles. *A Treatise on the Continued Fevers of Great Britain*. London: Parker, Son and Bourn, 1862.

Nuclein Therapy: Its Rationale, Methods and Results. Parke, Davis and Company Press, 1900.

Palmer, Love, and Henry Frieze. *Memorial of Alonzo Benjamin Palmer*. Cambridge, MA: Riverside Press, 1890.

Pasquier, Etienne. *Recherches de la France*. (*Research in French History*). Paris, 1596.

Patterson, K. David. *Pandemic Influenza: 1700–1900*. Totowa, NJ: Rowman and Littlefield, 1986.

Poisonous Proteins: The Herter Lectures for 1916. St. Louis, MO: C.V. Mosby, 1917.

Raoult, D., and M. Drancourt (eds.). *Paleomicrobiology: Past Human Infections*. Berlin: Springer-Verlag, 2008.

Reed, Walter, Victor Clarence Vaughan, and Edward Oram Shakespeare. *Report on the Origin and Spread of Typhoid Fever in United States Military Camps During the Spanish War of 1898*. Washington, D.C.: Government Printing Office, 1904.

Review of War Surgery and Medicine, Vol. 2 (1). Washington: Office of the Surgeon General, 1919.

Shaw, Wilfred (ed.). *The University of Michigan: An Encyclopedic Survey*, Vol. 2. Ann Arbor: University of Michigan Press, 1951.

Silverstein, Arthur. *A History of Immunology*. 2nd ed. New York: Elsevier, 2009.

Sisley, Richard. *Epidemic Influenza: Notes on its Origin and Method of Spread*. London: Longmans, Green, 1891.

Stedman, Thomas (ed.). *Twentieth Century Practice: Infectious Diseases*, Vol. 8 New York: William Wood, 1898.

Stedman, Thomas. (ed.). *Twentieth Century Practice of Modern Medical Science: Infectious Diseases*, Vol. 15. New York: William Wood, 1898.

Swithinbank, Harold, and George Newman. *Bacteriology of Milk*. New York: E.P. Dutton, 1903.*Thirteenth Annual Report of the Secretary of the State Board of Health of the State of Michigan for the Fiscal Year Ending September 30, 1885*. Lansing, MI: W.S. George, 1886.

Thompson, Theophilus (ed.). *Annals of Influenza or Epidemic Catarrhal Fever in Great Britain from 1510 to 1837*. London: Sydenham Society, 1852.

Thucydides' Peloponnesian War Book II.

Twelfth Annual Report of the Secretary of the State Board of Health of the State of Michigan for the Fiscal Year Ending September 30, 1884. Lansing, MI: W.S. George, 1885. January 8, 1884.

Twenty-Ninth Annual Report of the Secretary of the State Board of Health of the State of Michigan for the Fiscal Year Ending June 30, 1901. Lansing, MI: Wynkoop Hallenbeck Crawford, 1902.

University of Michigan, Board of Regents, *Regents' Proceedings: 1837–1864*. Ann Arbor: University of Michigan.

Vaughan, Victor C. *A Doctor's Memories*. Indianapolis: Bobbs-Merrill, 1926.

Vaughan, Victor C., and Frederick G. Novy. *Ptomaines and Leucomaines, or the Putrefactive and Physiological Alkaloids*. Philadelphia: Lea Brothers, 1888.

Vaughan, Victor C., and Frederick G. Novy. *Ptomaines, Leucomaines, Toxins and Antitoxins or The Chemical Factors in the Causation of Disease*. 3rd ed. Philadelphia: Lea Brothers, 1896.

Vaughan, Victor Clarence. *Epidemiology and Public Health: A Text and Reference Book for Physicians*, Vol. 1. St. Louis: C.V. Mosby, 1922.

Vaughan, Victor Clarence, and Frederick G. Novy. *Cellular Toxins, or, the Chemical Factors in the Causation of Disease*. 4th ed. Philadelphia: Lea Brothers, 1902.

Webber, Harry E. *Twelve Months with the Eighth Massachusetts Infantry in the Service of the United States*. Newcomb and Gauss, 1908.

Webster, Noah. *A Brief History of Epidemic and Pestilential Diseases*, Vol. 1. Hartford, CT: Hudson and Goodwin, 1799.

Whitaker, Catherine J. *The Early Years of the University of Michigan Medical School: The Upjohn Family Experience*. Ann Arbor: Michigan Historical Collections/Bentley Historical Library, 1982.

Wintermute, Bobby. *Public Health and the U.S. Military: A History of the Army Medical Department, 1818–1917*. New York: Routledge, 2011.

Yerkes, Robert Mearns (ed.) *The New World of Science: Its Development During the War*. New York, NY: Century, 1920.

Zinsser, Hans. *Infection and Resistance*. 2nd ed. New York: Macmillan, 1922.

Periodicals

American Journal of Hygiene
The American Journal of Public Health
Annals of Internal Medicine
Annals of Medical History
The Boston Medical and Surgical Journal
British Medical Journal
Cellular Immunology
Chemical Reviews
Clinical Infectious Diseases
Deutsche Medicinische Wochenschrift
Engineering News and American Railway Journal
European Journal of Epidemiology
First Annual Report of the Council of National Defense. Washington, 1917.

Fortschritte der Medizin
History and Philosophy of the Life Sciences
Infectious Disease Clinics of North America
Journal of Experimental Medicine
The Journal of Immunology
Journal of Infectious Diseases
The Journal of Laboratory and Clinical Medicine
Journal of Physiology and Pharmacology
The Journal of the American Medical Association
Journal of the History of Medicine and Allied Sciences
Journal of the Medical Society of New Jersey
Journal of the Senate of the State of Michigan, 1887
The Lancet
Medical Library and Historical Journal
The Medical News
The Medical World
The Michigan Alumnus
Mittheilungen aus den Kaiserlichen Gesundheitsamte
Nature
New York State Journal of Medicine
The New York Times
Northwestern Lancet
Official Bulletin, United States Committee on Public Information
Pediatric Nutrition
The Physician and Surgeon
Proceedings of the American Chemical Society
Proceedings of the American Philosophical Society
Progressive Medicine
Public Health, Michigan
Report of the National Academy of Sciences Fiscal Year 1924–1925
Revue d'Immunologie
Revue Scientifique et Technique
Science
Transactions of the Association of American Physicians
Transactions of the Clinical Society of the University of Michigan.
Zeitschrift für Hygiene

Web Sites

http://collections.nlm.nih.gov/ocr/nlm:nlmuid-101467619-bk
http://coursesa.matrix.msu.edu/~hst425/readings/Berliner.htm
http://elane.stanford.edu/wilson/html/chap22/chap22-sect3.html
http://english.byu.edu/facultysyllabi/KLawrence/WEBSTER.briefhistory.pdf
http://history.amedd.army.mil/booksdocs/wwii/PM4/CH22.Typhoid.htm
http://quod.lib.umich.edu/u/umregproc/ACW7513.1886.001/562
http://www.doctorsreview.com/history/nov05-history/
http://www.mocavo.com/Albert-Gitchell-1890-1968-Social-Security-Death-Index/16667150139699329692
http://www.nature.com/nature/journal/vaop/ncurrent/full/nature12377.html
http://www.ncbi.nlm.nih.gov/pmc/articles/PMC195158/?page=1
http://www.ncbi.nlm.nih.gov/pmc/articles/PMC2862337/
http://www.ncbi.nlm.nih.gov/pmc/articles/PMC2909426/
http://www.unboundmedicine.com/medline/citation/4944138/%5BFrancesco_Selmi_and_the_discovery_of_ptomaine__1870_%5D_
http://www7.nationalacademies.org/ocga/other/Act_to_incorporate.asp
Tarolli, Janet. "First Ladies at Michigan in Medicine." Medicine at Michigan (Fall 2000) 2(3); http://www.medicineatmichigan.org/magazine/2000/fall/women/

Index

Numbers in ***bold italics*** indicate pages with photographs.

Agramonte, Aristides ***68***
Albany Medical School 8
Allen, Dr. Jonathan Adams, Jr. 11
American Medical Association (AMA) 1, 9–10, 13–14, 20, 89, 92, 185
Andrewes, Sir Christopher 164
Arrowsmith (book) 45
Association of American Physicians 1, 40
Auer, Dr. John 183
Avery, Oswald 162, 173–174

Baker, Secretary of War Newton 138
Bevan, Dr. Arthur Dean 89–90
Bitter, Ludwig 61
Blackwell, Dr. Elizabeth 17
Blue, Dr. Rupurt 135, ***135***
Boivin, Andre 99
Bollinger, Dr. Otto 99–100
Bordet, Dr. Jules 52
Braisted, Dr. William 134, ***135***
Brieger, Dr. Ludwig 60–61, 64
Browicz, Tadeusz 85, 195n49
Bryan, William Jennings 76
Buchner, Dr. Hans 103–104
Budd, Dr. William 84–85, 195n25

Call, Dr. Emma Louisa 17, 19
Carroll, Dr. James ***68***, 74
Chicago World's Fair 48
Cole, Dr. Rufus 126, 171–172, 176
Conklin, Dr. Edwin 130
Corbin, Henry 195n10
Council of National Defense 131–132, 134, 171
Councilman, Dr. William 89–90

Cuba 4, 66, 69–70; *see also* Spanish-American War

Davaine, Dr. Casimir 58
Davenport, Charles 128
Denton, Dr. Samuel 11, 14
Detroit Medical College 125
d'Herelle, Dr. Felix 186–187
Dock, Dr. George 45, 50, 54–55, 74, 125–126
Douglas, Dr. Silas 15, 17, ***18***, 25–26, 28
Drake, Dr. Daniel 200n10
Dunglison, Dr. Robley 153
Dunning, Richard 158
Durham, Dr. Edward 86–87

Eberth, Dr. Karl 40, 85–86, 193n27
Ehrlich, Dr. Paul 50, 102–103, 106–108, 122–123
Escherich, Theodor 63, 194n25

Farnsworth, Elon 9
Flexner, Dr. Abraham 91, 185
Flexner, Dr. Simon 91, 130
Flugge, Dr. Carl 96
Ford, Dr. Coryden 15, ***18***, 21–22, 43
Fraenkel, Dr. Carl 2, 37
Frazier, Dr. Charles 89–90
Funston, General Frederick 201n98

Gaffky, Dr. Georg 85
Gamaleia, Nikolai 124
Geneva Medical College 15, 17
Gerhard, Dr. William 83
Gerow, Dr. Elizabeth Hait 17, 19
Gibbes, Dr. Heneage 44

Gitchell, Albert 166–167
Gluge, Dr. Gottleib 145–146, 151, 153
Goddard, Henry 128
Gone with the Wind (film) 137
Gorgas, Gen. William 131–132, 134, *135*, 135–138, 171
Gram, Hans Christian 95–96, 174
Gray, Dr. Asa 7, 8, 52
Grayson, Rear Adm. Cary *135*, 135–136
Great War (World War I) 4, 12
Grűnbaum, Dr. Albert 86–88
Gunn, Dr. Moses 11, 15

Hale, George Ellery 130–131
Halsted, Dr. William 192$ch3n$18
Hamilton, Dr. Alice 46, 79
Harrington, Dr. Mark 26
Hewlett, Dr. Albert Walter 125–127
Heyfelder, Dr. Johann 160
Hilgard, Dr. Eugene 26
Hippocrates 143
Hirsch, Dr. August 144–145, 153
Hoffa, Dr. Albert 59, 64, 194$ch5n$9
Houghton, Dr. Douglass 8
Hubbard, William 154
Huxham, John 157
Hygienic Laboratory (Ann Arbor) 2, 31, 37–39, 50, 113, 122

influenza 134, 138; and the American army 165; appearance in the Americas 154; eighteenth and nineteenth centuries 155–159; epidemic of 1510 147–148; epidemic of 1557 149–150; pandemic of 1580 151–153; pre-sixteenth century 143–147; structure of virus 139–140

Jefferson, Pres. Thomas 153
Jenner, Dr. Edward 158
Jenner, Sir William 83
Johns Hopkins University Medical School 12, 14, 40, 46, 73–74, 91, 125–126

Kektoen, Dr. Ludvig 185
Kelly, Dr. Howard 192$ch3n$18
Kitasato, Shibasaburo 34, 100
Klebs, Dr. Edwin 85
Koch, Dr. Robert 2, 34, *35*, 36–37, 39, 40; postulates 57, 164; 58, 60–61, 73, 85, 90, 97, 100, 105, 108, 116, 160
Koen, Dr. John 163

Langley, Dr. John 38
Lazear, Dr. Jesse *68*
leucomaines 3, 56, 58, 65, 109, 111, 186

Lewis, Dr. Paul 163, 183
Livy (Titus Livius) 143–144
Löeffler, Dr. Friedrich 61
Louis, Pierre Charles Alexandre *34*, 36, 82–83
HMS *Lusitania* 130

USS *Maine* 66
Mallon, Mary ("Typhoid Mary") 42, 198$ch9n$10
Martin, Dr. Edward 135, *135*
Martin, Dr. Franklin 134, 136
Mason, Stevens (governor) 7
Mayo, Dr. Charles *135*, 135–136
Mayo, Dr. William 136
McClintock, Charles 50–52
McKinley, President William 66
Mesrobeanu, Lydia 99
Metchnikoff, Dr. Elie 103, 106–108
Michigan State Board of Health 1, 2, 30–31, 37
Minor, Dr. Loring 167
Mirsky, Dr. Alfred 180
Mosher, Dr. Eliza Maria 17, 19
Mount Pleasant College 1, 5
Mundy, Edward 9

National Academy of Sciences 129, 131
National Research Council 4, 129–133, 180, 185
Nicolaier, Dr. Arthur 96
Nicolle, Dr. Charles 116
Novy, Frederick G. 3, 6, 32–34, 37–38, 44–45, *56*, 57, 60, 93, 98, 100–102, 104, 109, 112
Nuttall, Dr. George 102

Osler, Dr. William 12–14, 40, 45, 50, 55, 74, 192$ch3n$18

Palmer, Dr. Alonzo 14, 17, *18*, *19*, 21, 29, 43
Panum, Peter 60
Pasquier, Etienne 146–147
Pasteur institute (Ann Arbor) 110–111
Pfeiffer, Dr. Richard 86, 97–98, 160, 162–163, 169, 174–176
Pitcher, Dr. Zina 7, 9, 10, 15
Plague of Athens 6, 81–82
Prescott, Dr. Albert 25–26, 38, 53
ptomaines 3, 32, 56, 58–61, 64, 99, 111, 186, 192$ch3n$8

Reed, Dr. Walter 4, 47, *68*, 72–73, 76–77, 79, 89
Rensselaer Polytechnic Institute 8

Richard, Dr. Charles 173-174
Richet, Dr. Charles 197n73
Roos, C.G.A. 162
Rose, Dr. Preston 26-28
Roux, Emile 61
Rush, Dr. Benjamin 200n67
Russell, Dr. Frederick 171

Sager, Dr. Abram 8, 11, 14-15, *18*
Sanford, Dr. Amanda 17-18
Searing, Dr. Anna Hutchinson 17, 19
Selmi, Francesco 60, 64
Sewall, Dr. Henry 33, *34*, 43
Shakespeare, Dr. Edward 4, 47, 72-73, 76-77, 79, 89
Shope, Dr. Richard 163-164
Short, Dr. Thomas 147-150, 152, 155
Simpson, Dr. Frank 134, *135*
Smith, Dr. Wilson 164
Snow, Dr. John 181
Soper, Dr. George 138, 175-177, 198ch9n10
Spanish-American War 4, 6, 47, 52, 66, 85, 132, 178, 181
Sternberg, Dr. George 67, 71, *72*, 73, 75
Stockwell, Cyrus 16

typho-malaria 47, 70, 74, 80, 88; see also typhoid
Typhoid Commission 70-80, 88, 186
typhoid fever 4, 47-49, 66, 70, 71, 73, 76, 77, 78; history 80-88; 115, 118-119, 143, 161, 178, 180; see also Typhoid Commission
tyrotoxicon 1, 32, 56, 62, 64

University of Michigan 1, 3, 5, 6-7, 14-15, 91; curriculum 21, 22, 39, 45-47, 89-90; establishment of medical school 10-11

Vaughan, Henry Frieze (son) 3
Vaughan, Dr. Victor Clarence 47, *71*, *94*, *135*, *172*, *187*; ancestry 23-24; appointed dean of medicine 43; appointed Medical Corp 136; and Chicago World's Fair 48; death 189; eugenics 128; graduate education 25-28, *29*; implementation of "Michigan Method of water analysis 113-115; lecturer 29; professor 38; resignation as dean 179-180; secondary education 24-25; service in Spanish-American War 69-80; see also National Research Council
Vaughan, Warren Taylor (son) 145, 186
Vaughan tank 111-113
Victor Vaughan Student Society 4
Vines, Richard 154
von Behring, Dr. Emil 34, 50, 96, 102, 108
von Grűber, Dr. Max 86-87
von Pettenkofer, Dr. Max 180-181

Walcott, Charles 130
Webster, Noah 154-155, 200n67
Welch, Dr. William 12, 36, 40, 41-42, 130-131, 134-135, *135*, 137, 171-173, 176
Wheeler, May 118, 122
Widal, Dr. Fernand 71, 86, 87
Wilson, Pres. Woodrow 129-131, 180
Winthrop, John 154
Witherspoon, Dr. J.A. 89-90
Wolbach, Dr. Simeon Burt 173, 175
Woodward, Robert 130
World War I (The Great War) 129, 134

USS *Yale* 69
Yersin, Dr. Alexandre 61
Ypsilanti (Michigan) 110

Zinsser, Dr. Hans 97-98

www.ingramcontent.com/pod-product-compliance
Ingram Content Group UK Ltd.
Pitfield, Milton Keynes, MK11 3LW, UK
UKHW042001140426
5217IPUK00015B/924